AF589381

NANOBIOSENSOR
CONCEPTS & APPLICATIONS

THE AUTHOR

Dr. S. Thirumalairajan, working in the most prestigious DBT-Ramalingaswami Re-entry Faculty Fellow, India in the Centre for Agricultural Nanotechnology, Coimbatore, India. He received a Ph.D. in Physics-Nanoscience and Technology (Interdisciplinary) from Bharathiar University, India. He is an expert in the fields of functional nanostructures materials for sensors, energy, food safety applications. He has received two externally funded projects with a total cost of ~ Rs. 2 crores to address issues for food safety applications and also received recognition as a DST-SERB-PDF (Selected),CSIR- SRF, Jawaharlal Nehru Doctoral Studies, URF,Govt. of India. He hasvisited and worked as a senior postdoctoral fellow,post-doctoral fellow, and guest scientist in various countries like Brazil, USA, France, South Korea, Indonesia, and Taiwan, for the past 10 years and the outcome of novel research resulted in published India patent (4), UK design patent grant (1), high impact international journals (34) national journals (2), book (4), book chapters (6), conferences presentation (25), and received 10 research awards and distinctions for his credentials. He is a member of various scientific bodies inthe national and international community.

Dr. K. Girija is an Assistant Professor of Physics and i/C Head of the physics department, at Dr. N.G.P. Arts and Science College, Coimbatore. She completed doctor's degree in Physics - Nanoscience and Technology from the Department of Nanoscience and Technology, Bharathiar University, India. Her research also addresses the fabrication and evaluation of nanomaterials for use in multifunctional applications. Dr. K. Girija has received recognition as a Post-Doctoral Fellow from FAPSEP, Brazil,Senior Research Fellow, CSIR, Government of India and Junior Research Fellowship from DRDO - BU CLS, Coimbatore. She has published 27 articles, 1 UK design patent grant, 2 Indian patents published, 3 books, 8 book chapters, and guides post graduate and Ph.D. scholars. She is a member of the Materials Research Society, Fellow of Asian Research Association, Electron Microscope Society and Indian Society for Technical Education.

NANOBIOSENSOR
CONCEPTS & APPLICATIONS

– Authors –

S. Thirumalairajan
K. Girija

2025

Scholars World
A Division of
Astral International Pvt. Ltd.
New Delhi – 110 002

ISBN: 9789359193953

Published by : **Scholars World**
A *Division of*
Astral International Pvt. Ltd.
– ISO 9001:2015 Certified Company
4736/23, Ansari Road, Darya Ganj
New Delhi-110 002
Ph. 011-43549197, 23278134
E-mail: info@astralint.com
Website: www.astralint.com

PREFACE

The short form of the biological sensor is known as a biosensor. The bio-element communicates through the analytes being checked and the biological reply can be changed into an electrical signal using the transducer. Biosensors can be defined as analytical devices, which include a combination of biological detecting elements like a sensor system and a transducer. In this sensor, a biological element is an enzyme, a nucleic acid otherwise an antibody. When we compare with any other presently existing diagnostic device, these sensors are advanced in the conditions of selectivity as well as specificity. The applications of Biosensors mainly include checking ecological pollution control, in the agriculture field as well as food industries. The main features of biosensors are stability, cost effect, sensitivity, and reproducibility. In the First type of biosensor, the reaction of the product disperses to the sensor and causes an electrical reaction. In the second type, the sensor involves particular mediators between the sensor and the response in order to produce a better response. In the third type, the response itself causes the reaction and no mediator is straight involved.

From both the scientific and technological points of view, it is worthwhile to learn the ways to fundamental of the biosensor, characteristic behaviour, different type of biosensors and its functional properties towards various commercial applications are crucial and essential. The authors discuss in this present book the fundamental understanding of biosensors, electrochemical biosensors, optical, thermal sensors, bio-recognition, transducers, and agricultural applications.

In this book consists of five chapters and the contents of each chapter are as follows: the first chapter gives an introduction of nanoscience and technology, sensors, characteristic, and classification of sensing parameters. The second chapter describes in detail the fundamental of electrochemical biosensor, principles of electrochemical sensor and their techniques. The third chapter explains the principle of optical, colorimetric sensor such as conductometry, voltammetry, Impendence biosensor, and glucose biosensor. The fourth chapter deals with the transducers and bio-recognition elements such as enzymes, antibodies, receptors, nucleic acid and immobilization of biomolecules. It is worth to note that the different applications of biosensor in agriculture as discussed in the fifth chapter.

Dr S. Thirumalairajan
Dr. K. Girija

CONTENTS

Chapter 1

BASIC OF SENSOR

1.1 Introduction to Nanoscience and Technology

Nanoscience is an emerging area of science, which concerns itself with the study of materials that have very small dimensions, in the range of nanoscale. The word itself is a combination of nano, from the Greek "nanos" (or Latin "nanus") meaning "Dwarf", and the word "Science" meaning knowledge. It is an interdisciplinary field that seeks to bring about mature nanotechnology, focusing on the nanoscale intersection of fields such as physics, biology, engineering, chemistry, computer science, and more. Nanoscience is the study of phenomena on a nanometer scale means 10^{-9} m. Atoms are a few tenths of a nanometer in diameter and molecules are typically a few nanometers in size. A nanometer is a magical point on the length scale, for this is the point where the smallest man-made devices meet the atoms and molecules of the natural world. Therefore, a nanometer is one billionth of a meter and it is the unit of length that is generally most appropriate for describing the size of single molecule. Nanometer objects are too small to be seen with naked eye. Infect, if one wanted to see a 10 nm sized marble in his hand, his eye would have to be smaller than a human hair. Anyhow the rough definition of nanoscience could be anything which has at least one dimension less than 100 nanometres.

1.1.1 Classification of Nanomaterials

Nanomaterials can be classified dimension as shown in Figure 1.1 into following categories.

- Zero Dimensional < 100 nm nanosphere, Quantum dots, cluster
- One dimension < 100 nm nanorods, nanowires, etc.,

- Two dimensions < 100 nm Tubes, fibers, platelets, etc.,
- Three dimensions < 100 nm particles, hollow Spheres, etc.

On the basis of phase composition, nanomaterials in different phases can be classified as,

- Single phase solids include crystalline, amorphous particles and layers, etc.
- Multi-phase solids include matrix composites, coated particles, etc.

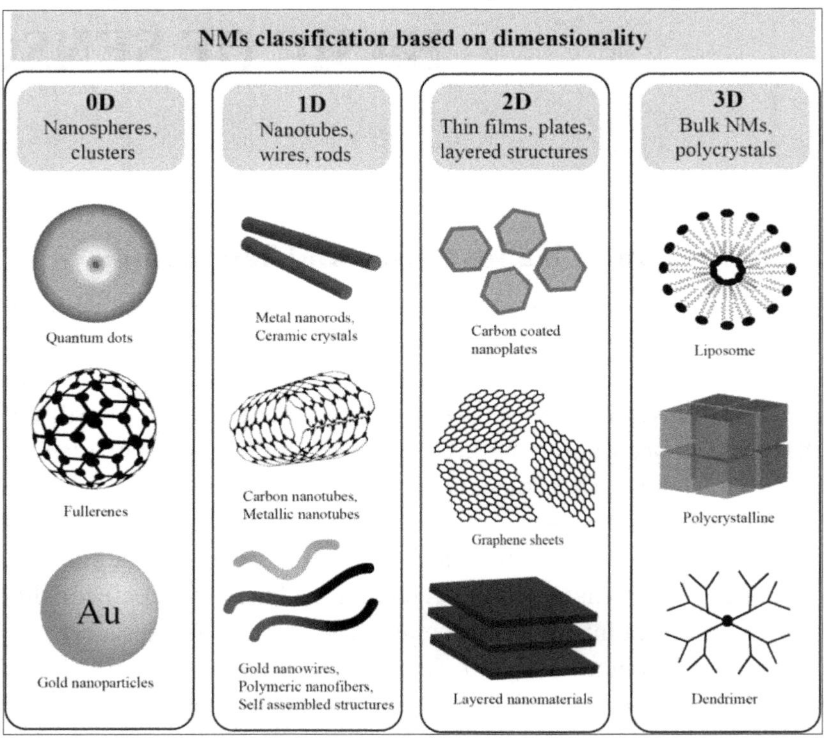

Figure 1.1 Schematic illustrating the relative dimensions of nanoparticles with

1.1.2 Nanoscience and Physics

Physics is the mother of natural sciences. In principle, physics can be used to explain everything that goes on at the nanoscale. There is active physics research going on in nano mechanics, quantum computation, quantum teleportation, artificial atoms etc. At nanometer scale physics is different. Properties not seen on a macroscopic scale now become important- such as quantum mechanical and thermodynamic properties. Rather than working with bulk materials, one works with individual atoms and molecules. By learning about an individual molecule's properties, we can put them together in very well-defined ways to produce new materials with new and amazing characteristics. Some Physical properties of materials reduced to nano scale can suddenly show very different properties compared to what they exhibit on a macro scale, enabling unique applications.

For instance:

- Copper, which is an opaque substance, become transparent.
- Platinum, which is an inert material, becomes the catalyst.
- Aluminium, which is a stable material, turns combustible.
- Silicon insulators become conductors.
- Gold, which is solid, inert, and yellow on room temperature at microscale, becomes liquid and red in colour at the nanoscale on room temperature.

1.1.3 Nanotechnology

The nanoworld provides scientists with a rich set of materials useful for probing the fundamental nature of matter. These materials have unique structures and tunable properties. This makes them valuable for many different real-world applications nanotechnology is currently in a very infantile stage. It is basically the creation of Useful functional materials, devices and systems through control of matter on the nanometer length scale and exploitation of novel phenomena and properties (physical, chemical, biological) at that length scale.

1.1.4 Applications of Nanotechnology

The potential applications of nanotechnology in different fields are the following:

- Electronics
- Health and Medicine
- Energy and Environment
- Space exploration
- Agricultural science
- Food technology

1.2 Chemical Sensor

A sensor is a device that receives a signal or stimulus and responds to the stimulus in the form of an electrical signal. The output signals correspond to some forms of electrical signal, such as current or voltage. The sensor receives different kinds of signals *i.e.*, physical, chemical, or biological signals, and converts them into an electric signal. It is classified into different types based on the applications, input signal, and conversion mechanism.

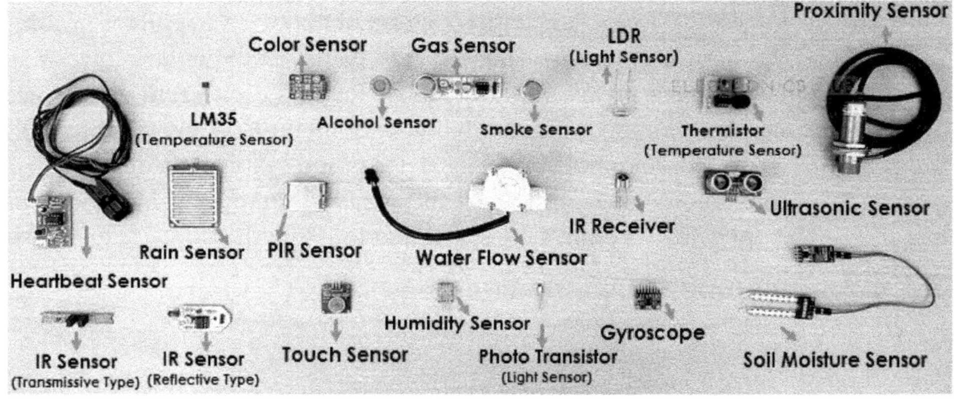

Figure 1.2 Photograph image of Different types of sensor (Courtesy: From Web)

There are two main types of sensors: passive sensor and active sensor. A passive sensor does not require any extra energy source and electric signal is produce directly reply to stimulus of external sources. This means that the sensor converts input energy to output signal energy. Examples of passive sensors include photographic, thermal, electric field g, chemical, infrared and seismic sensin. The active sensors need external sources of energy for their response, known as excitation signal. To produce the output signals, sensors adopt necessary changes to these input signals. The active sensors are also known as parametric sensors due to their own properties, which can be modified in response to an exterior effect, and these properties can be afterward changed into electric signals. Active sensors have a variety of applications related to meteorology and observation of the Earth's surface and atmosphere.

1.2.1 Passive Sensors vs. Active Sensors:

S.No	Passive Sensors	Active Sensors
1	Does not require external power	It requires external power
2	It can only be used to detect energy when the naturally occurring energy is available	Provides its own energy source for illumination
3	No interference problem in the environment	Less interference problem
4	Can operate in the same environmental conditions	Can operate in different environmental conditions
5	It is sensible for weather conditions	It's not sensible for weather conditions
6	Not well suited for dark conditions	Well suited for dark conditions
7	Difficulties in interpreting the output signals	Easy to interpret the output signals
8	Less control of noise	Better control of noise
9	Low cost	High cost
10	Ex: Camera, sonar	Ex : LASER ,Radar

1.2.2 Sensors based on their detection properties:

S.No	Types	Properties	Sample Images
1	Thermal Sensor	Temperature, Heat, Flow of Heat etc.	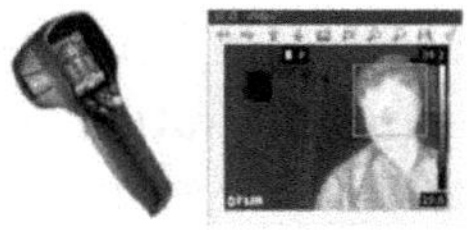
2	Electrical Sensor	Current, Voltage, Inductance, Resistance etc.	
3	Magnetic Sensor	Magnetic strength, mgnetic flux density etc.	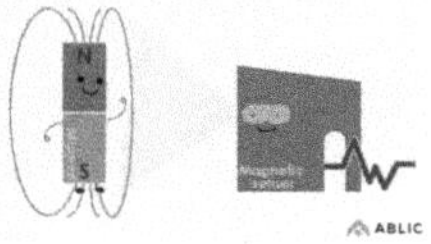
4	Optical Sensor	Intensity of Light, Wavelength, Polarazitation etc.	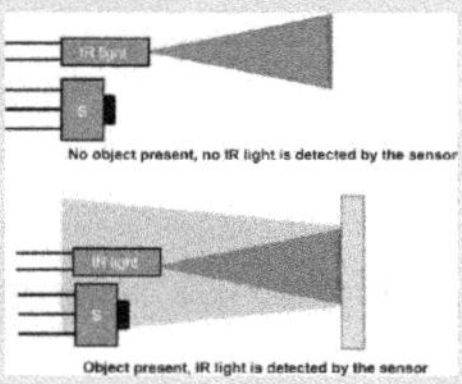

5	Chemical Sensor	Composition, pH, concentration etc.	
6	Pressure Sensor	Pressure, force etc.	
7	Vibration Sensor	Displacement, acceleration, velocity etc.	
8	Moisture Sensor	Water content, moisture level etc.	
9	Speed sensor	Acceleration, Velocity etc.	

1.2.3 Chemical Sensor Mechanism

The chemical sensor is an analyser that responds to a particular analyte in a selective and reversible way and transforms input chemical quantity, ranging from the concentration of a specific sample component to a total composition analysis, into an analytically electrical signal. The chemical information may originate from a chemical reaction by a biomaterial, chemical compound or combination of both attached onto the surface of a physical transducer towards the analyte as shown in Figure 1.3.

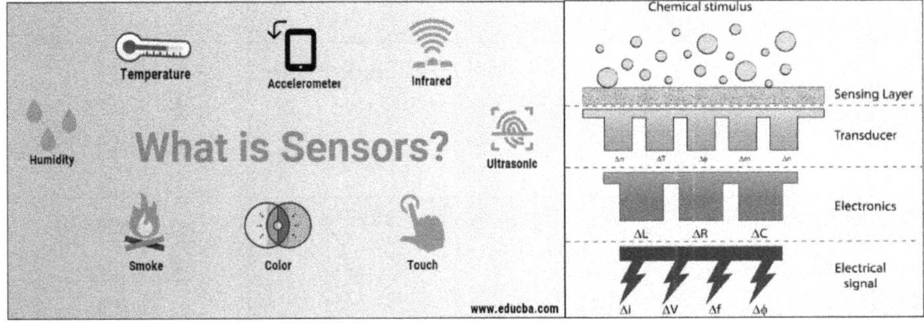

Figure 1.3 Schematic illustration of working principle of a chemical sensor (Courtesy: web)

According to the working principle, the chemical sensor can be classified into many types such as optical, electrochemical, mass, magnetic, and thermal as shown in Figure 1.4. The optical chemical sensor is based on the changes in optical phenomena analysis arising from the interaction between the analyte and the receiver. The electrochemical sensor utilizes an electrochemical effect among the analytes and featured electrodes. The working principle of the mass sensor depends on the quality change induced by the mass loading from the adsorption toward the analyte by the special modification of the sensor surface. The magnetic device is based on the magnetic properties in analyte adsorption, whereas the thermal sensor utilizes the thermal effect generated by the specific chemical reaction or adsorption process.

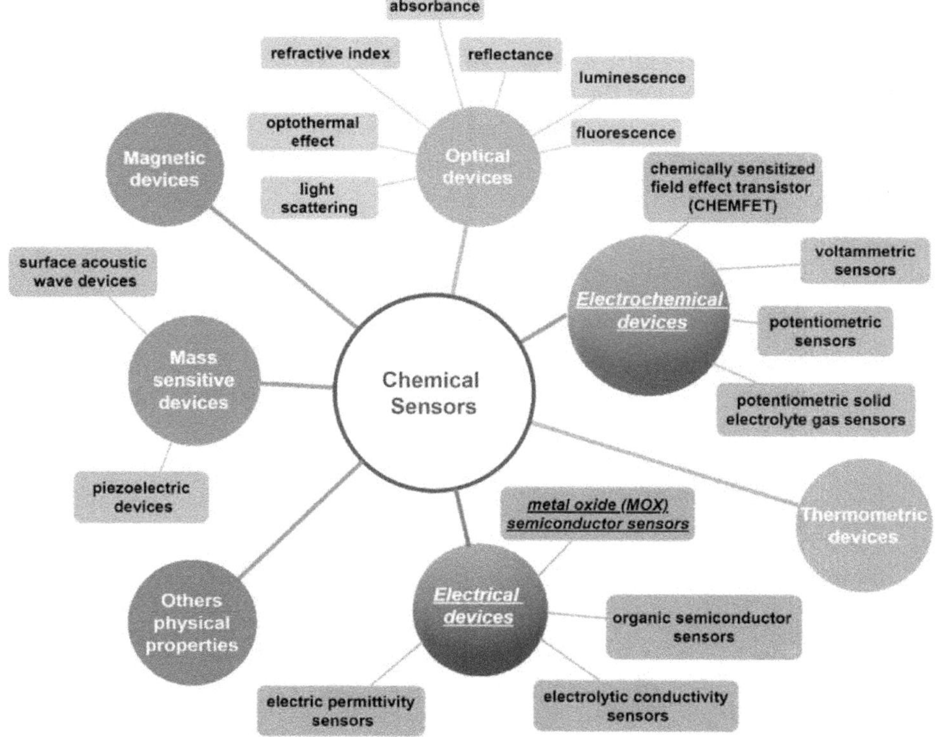

Figure 1.4 Classification of Chemical Sensors (Enza Fazio, MDPI, 3April 2021)

Another way to categorize the chemical sensors are based on the object to be detected, that is, the chemical sensors can be classified as gas sensors for trace gas analysis and monitoring, various sensors represented by the pH sensor, humidity sensor, and biosensors made by biological characteristics.

1.3 Biosensors

A biosensor is an analytical device, used for the detection of a chemical substance that combines a biological component with a physicochemical detector. The sensitive biological element, e.g. tissue, microorganisms, organelles, cell receptors, enzymes, antibodies, nucleic acids, etc., is a biologically derived material or biomimetic component that interacts with, binds with, or recognizes the analyte under study as shown in Figure 1.5. The transducer or the detector element, which transforms one form of signal into another form of a signal, works in a physicochemical way: optical, piezoelectric, electrochemical, electrochemical luminescence, etc., resulting from the interaction of the analyte with the biological element, to easily measure and quantify. The biosensor reader device connects with the associated electronics or signal processors that are primarily responsible for the display of the results in a user-friendly way. This sometimes accounts for the most expensive part of the sensor device, however it is possible to generate a user-friendly display that includes a transducer and sensitive element (holographic sensor). The readers are usually custom-designed and manufactured to suit the different working principles of biosensors.

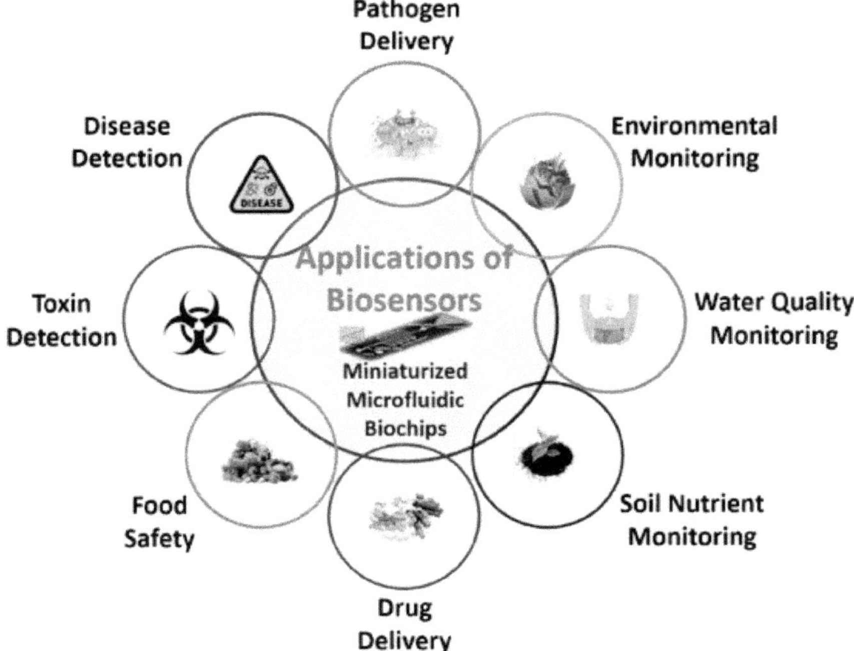

Figure 1.5 Schematic illustration representation that the applications of Bio-sensors

1.3.1 Working Principle

Biosensors works on the principle of signal transduction and bio-recognition of the elements. All the biological materials including enzymes, antibody, nucleic-acid, hormone organelle or whole cell can be used as sensors or detectors in a device. But the desired bio-receptor is usually a specific deactivated enzyme. The deactivated enzyme is placed in proximity to the transducer. The tested analyte links to the specific enzyme (bio-receptor) and inducing a change in the biochemical property of enzyme. The changes in-turn gives an electronic response through an electro-enzymatic approach. Electro enzymatic process is the chemical process of converting the enzymes into corresponding electrical signals with the aid of transducer. Now, the outcome from transducer i.e. electrical signal is a direct representation of the biological material being measured as shown in Figure 1.6. The electrical signal is usually converted into physical display for its proper analysis and representation.

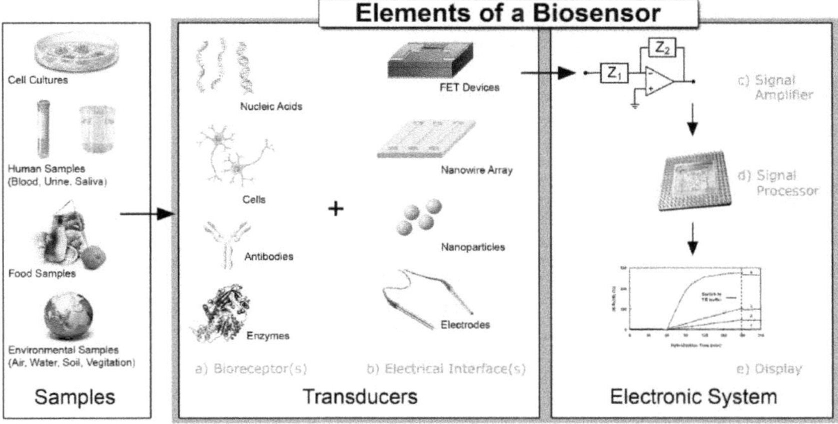

Figure 1.6 Schematic illustration of the working principle of a biosensor (*Dave, 30 March 2019*)

1.4 Classification of Sensors

The sensor device as well as the biological material of the biosensors are classified as:

- Electrochemical biosensors
- Calorimetric or thermal detection biosensors
- Optical biosensors
- Piezo-electric biosensors

- Resonant biosensors

1.4.1 Electrochemical Biosensor

Generally, electrochemical biosensor works on the principle that many enzyme catalysis reactions consume or generates ions or electrons causing some change in the electrical properties of the solution which can be detected and used as a measuring parameter. For example, some biological compounds such as glucose, urea, cholesterol, *etc.,*) are not electroactive, so the combination of reactions by this biosensor produces an electroactive element as shown in Figure 1.7. This electroactive element results in a change of current intensity, which is proportional to the concentration of analyte. An electrochemical biosensor uses an electrochemical cell with electrodes of different dimension and modifications.[5] Three kinds of electrodes are generally used-

- Working electrode
- Reference electrode
- Counter or Auxiliary electrode

It is the working electrode where reaction occurs between electrode substrate and analysis.

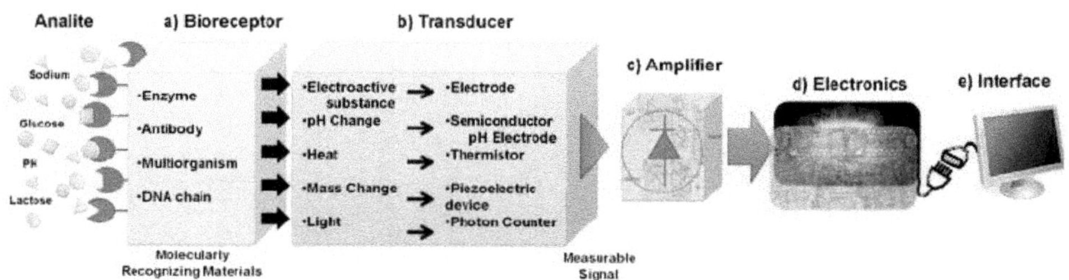

Figure 1.7 Working of Electrochemical Biosensor (Vikas Dull, 3 October 2013)

1.4.2 Calorimetric or Thermal Biosensor

Most of the enzyme-catalysed reactions are exothermic in nature. Calorimetric biosensors measure the change in temperature of analyte solution following enzyme action and interpret it in terms of analyte concentration in the solution. The analyte solution is passed through a small packed bed column consisting immobilized enzyme. The temperature of the solution is measured just before the entry of the solution into the column, and just as it leaves the column using separate thermistors as shown in Figure 1.18. It is the most usually applicable type of biosensor and can also be used for turbid and colourful solutions. There are demerits such as

- The biggest demerit is to maintain the temperature of the sample stream say + or - 0.01°C.
- Low range and sensitivity.

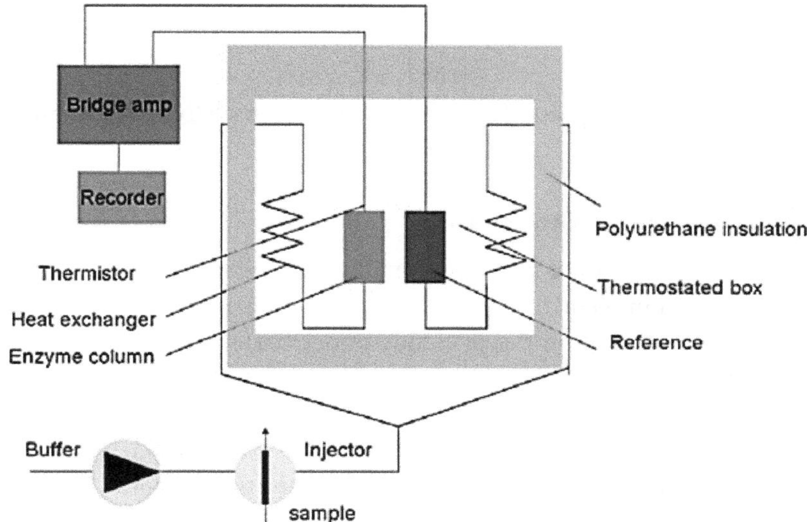

Figure 1.8 Schematic representation of the thermal Biosensor (*Yan Zhou, 6 November 2012*)

1.4.3 Optical Biosensor

Optical biosensors emit an optical signal, which is directly proportional to the concentration of the analyte. The bio-recognition elements used by biosensors are generally biological materials, including enzymes, antibodies, antigens, receptors, nucleic acids, whole cells, tissues. Both catalytic and affinity reactions are measured by this biosensor. The products generated during the catalytic reactions cause a change in fluorescence that is measured by the biosensor. In other way, biosensors measure the change induced in the intrinsic optical properties of the biosensor surface due to loading on it of di-electric molecules like protein as shown in Figure 1.19. A most advanced biosensor involving luminescence uses luciferase enzyme for the detection of bacteria in food or clinical samples. In the presence of O2, luciferase takes up the ATP released from the lysis of bacteria to produce light which is detected and measured by biosensor.

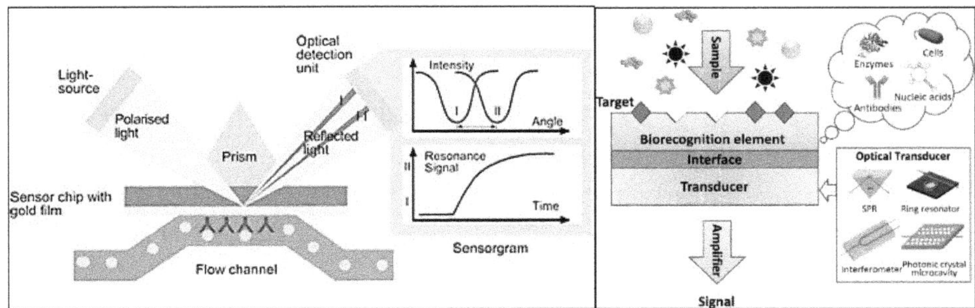

Figure 1.9 Schematic representation of optical biosensor (Nishtha Khansili, 15 July 2018)

1.4.4 Piezo-electric Biosensor

Piezoelectric biosensors are a group of analytical devices working on a principle of affinity interaction recording. A piezoelectric platform or piezoelectric crystal is a sensor part working on the principle of oscillations change due to a mass bound on the piezoelectric crystal surface. In these biosensors, the surface is coated with antibodies, which binds to the complementary antigen present in the sample solution.10,11 This result in increased mass, which decreases their vibrational frequency, this alteration/change, is used to determine the amount of antigen present in the sample solution as shown in Figure 1.10.

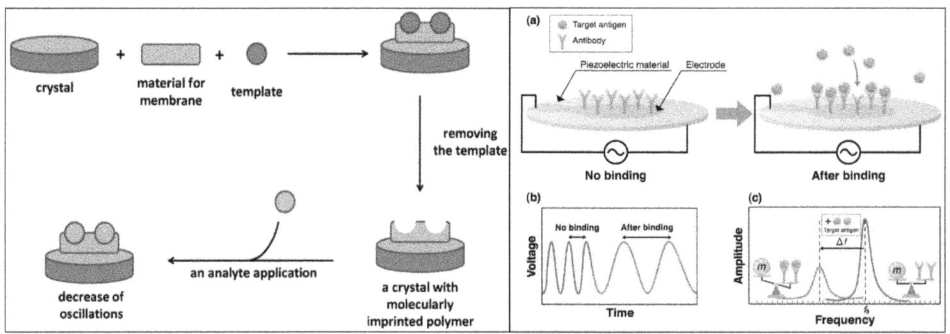

Figure 1.10 (a) Overview of Piezo-electric biosensor (Miroslav Pohanka, 19 March 2018) (b) Piezo-electric Biosensor (Fumio Narita, 24 November 2020)

1.4.5 Resonant Biosensor

The vibrations of the electron cloud in a molecule are termed as resonant biosensors. These Plasmon's oscillate at a particular frequency characteristic of the material and the oscillations in surface plasmon's are confined to the surface of the material as shown in Figure 1.11. Generally, gold or silver surfaces are preferred for the SPR based biosensors. When electromagnetic radiation falls on the metal surface, at a particular angle of incidence, the frequency of the electromagnetic radiation matches the frequency of vibrations resulting in resonance. The resonant angle depends on the refractive index of the medium. The local mass density on the metal surface in turn determines the refractive index. If the surface of the metal film is modified with the antibody/receptor i.e. capture molecule, then specific binding occurs between the capture molecule on addition of the sample and its ligand leading to an alteration in mass and hence change in resonant angle. These biosensors are employed to understand the functional aspects of human immune deficiency virus (HIV) both qualitatively and quantitatively. The major merits of these biosensors are rapid measurements and relatively high sensitivity. The major demerit is that it cannot be used to detect and measure the turbid and coloured solutions. In few cases, ligands may interfere with the binding.

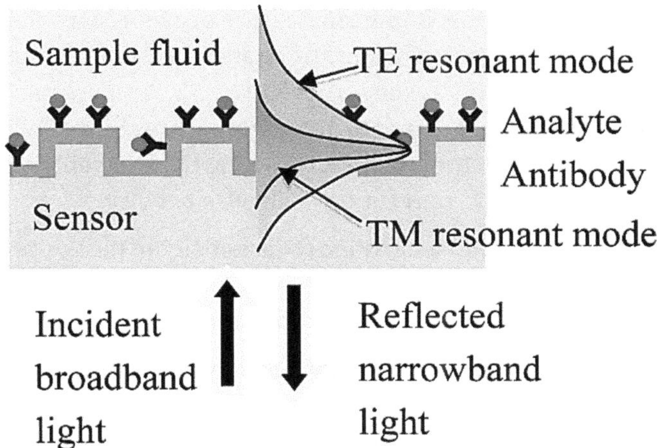

Figure 1.11 Schematic illustration of resonant Biosensor (*Robert Magnuson, 26 January 2011*)

1.5 Sensitivity

Sensitivity indicates how much the output changes when the input quantity it measures changes as shown in Figure 1.12. This parameter is sometimes confused with the detection limit. The biosensing sensitivity is improved by selectively immobilising the antibody of a specific region of high electromagnetic field intensity. We can visualize this effect by using atomic force microscopy and Electron microscopy. The sensitivity of a biosensor can be calculated by, dividing the slope of linearity graph by the active area of a biosensor. Peptide Nucleic Acid (PNA) biosensors are highly sensitive & are used for bio sensing purposes.

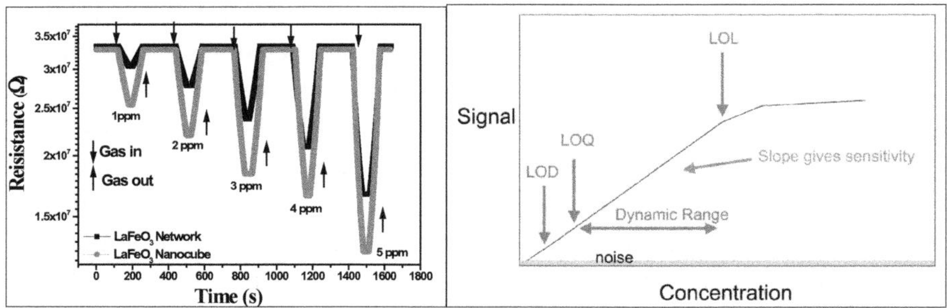

Figure 1.12 (a) sensor concentration and (b) the slope y/x represents the selectivity of a sensor

1.6 Selectivity

Selectivity is perhaps the most important feature of a biosensor. Selectivity is the ability of a bio receptor to detect a specific analyte in a sample containing other admixtures and contaminants as shown in Figure 1.13. It represents one of the key advantages of biosensors, compared to other methods, as they allow to determine an analyte in a complex mixture without resorting to prior separation. There are some simple solutions to circumvent the lack of selectivity:

- The biosensor could be destined to detect all the recognized compounds and provide the result as a global estimation of all substances present in the sample renouncing to the expectation as being selective.
- The usage of the biosensors is reduced only to samples that are known not the contain the potential interferents or that contain the analyte in huge excess in comparison with the expected level of interfering compounds or more complicated sample pre-treatments and purifications steps are carried out before the actual analysis with the biosensors.

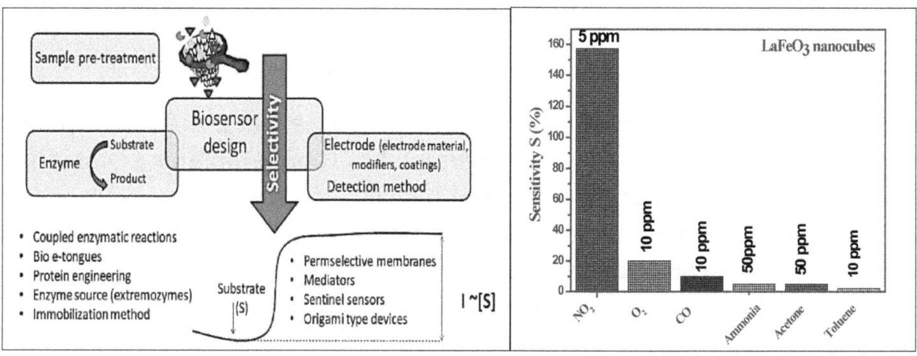

Figure 1.13 Schematic illustration of selectivity mechanism (b) characteristic of Selectivity of a Biosensor

1.7 Reproducibility

Reproducibility is the ability of a biosensor to generate identical responses for a duplicated experimental setup. The reproducibility of a sensor is characterised by the precision and accuracy of the transducer and electronics in a sensor [15]. It is the extent to which a tool is capable of producing the same result when used repeatedly in the same circumstances. The terms 'repeatability' and 'reliability' are often used synonymously with 'reproducibility as shown in Figure 1.14.

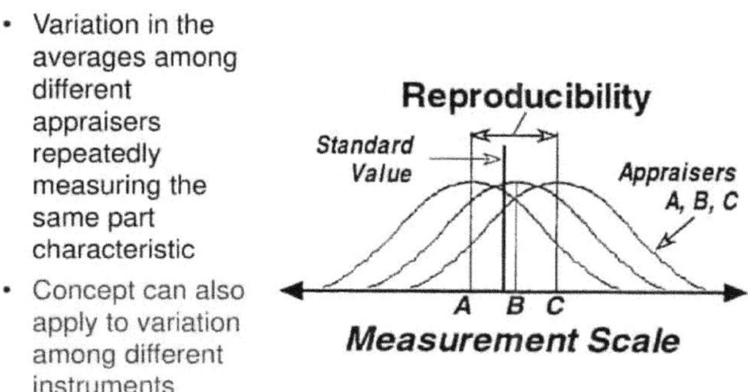

Figure 1.14 (a) Reproducibility of Biosensors (M. Mohiuddin, RH 2016)

1.8 Portability

The portability of biosensors for on-site diagnosis is limited due to various issues, including sample preparation techniques, fluid handling techniques, limited life time of biological reagents, device packaging, and integrating electronics for data collection/analysis and the requirement of external accessories and power. .

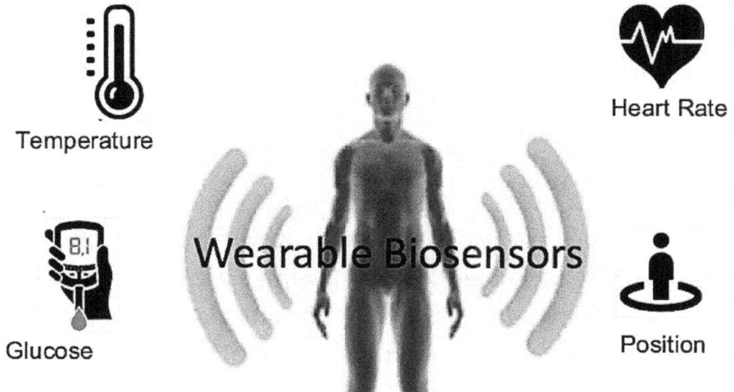

Figure 1.15 Portable (Wearable) Biosensors (Melis Kecili, 16 December 2019)

A recent development in portable biosensors allows rapid, accurate, and on-site detection of biomarkers, which helps to prevent disease spread by the control of sources. Less invasive sample collection is necessary to use portable biosensors in remote environments for accurate on-site diagnostics and testing as shown in Figure 1.15

1.9 Stability

Stability is the ability of a sensor to provide reproducible results for a certain period of time. This includes retaining the sensitivity, selectivity, response, and recovery time. It is the degree of susceptibility to ambient disturbance in and around the biosensing system as shown in Figure 1.16. Degradation rate is linearly dependent on temperature and by utilizing the proposed models. Operational Stability may be defined as the retention of activity of a protein or enzyme when in use. This is often the most quoted parameter in biosensor publications and it relates to the operating lifetime and reusability of a device. In many cases the lifetime of use of a sensor can be important, usually in an environment where monitoring of an analyte is required or when an analytical device incorporates a reusable sensor in the measuring process, e.g. the Yellow Springs Instrument glucose analyser [16].

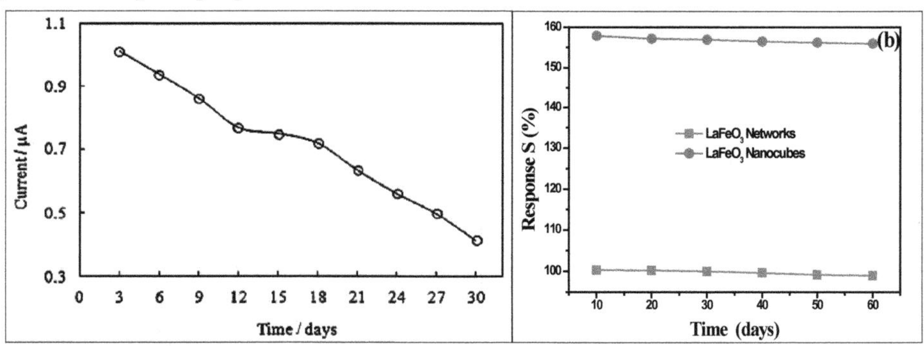

Figure 1.16 (a) Schematic illustration of mechanism (b) real sample analysis of stability

1.10 Detection Limit

The limit of detection (LOD) is defined as the lowest concentration of an analyte in a sample that can be consistently detected with a stated probability (typically at 95% certainty) as shown in Figure 1.17. The simple and very objective method is the determination through the repeatability of small concentrated sample measurement (for example the lowest calibration standard), LOD = 3*SD where, LOD = Limit of Detection and SD = Standard Deviation.

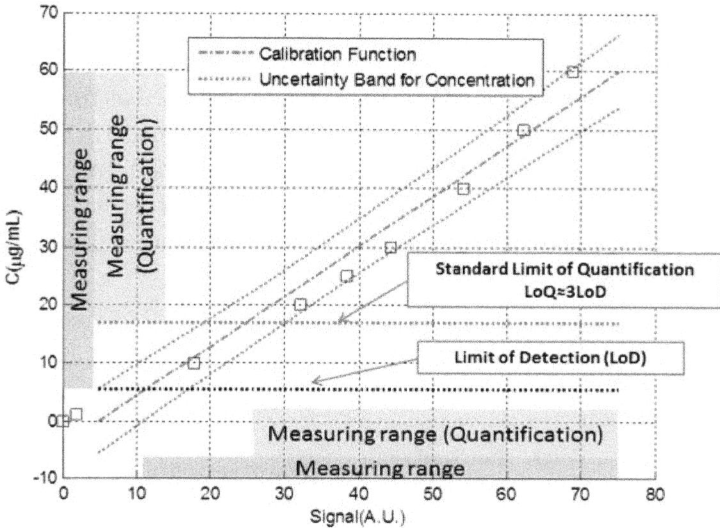

Figure 1.17 Schematic diagram representation that the detection Limit of a sensor (Alvaro Lavin, 2018)

1.11 Response Time

The biosensor response time is the time it takes for the sensor output to reach its new signal after being exposed to a change in analyte concentration, this is often quoted as T90, which is the time it takes for the signal to reach 90 % of its final value as shown in Figure 1.18.

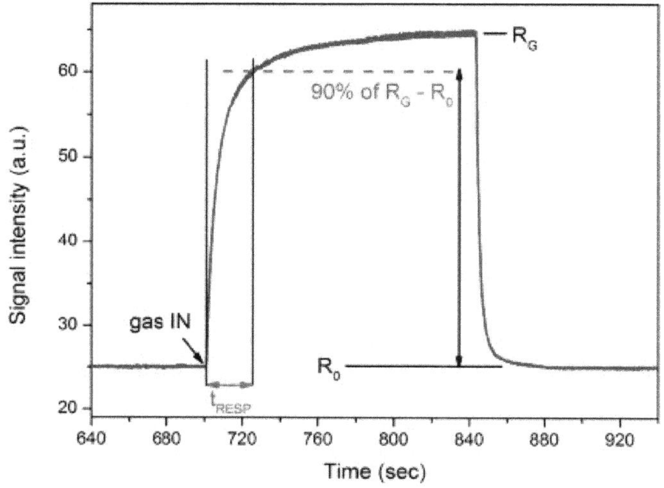

Figure 1.18 Response Time of Biosensors (Mushtaque Hussain, 2014)

1.12 Recovery Time

Recovery Time is defined as the time for a sensor to return to baseline value after the step removal of the measured variable. Usually specified as time to fall to 10% of final value after step removal of measured variable as shown in Figure 1.19. The recovery time objective (RTO) is the amount of time or real time during or after a disaster that can elapse without a business restoring its services and processes to acceptable levels before it will experience intolerable consequences associated with the disruption.

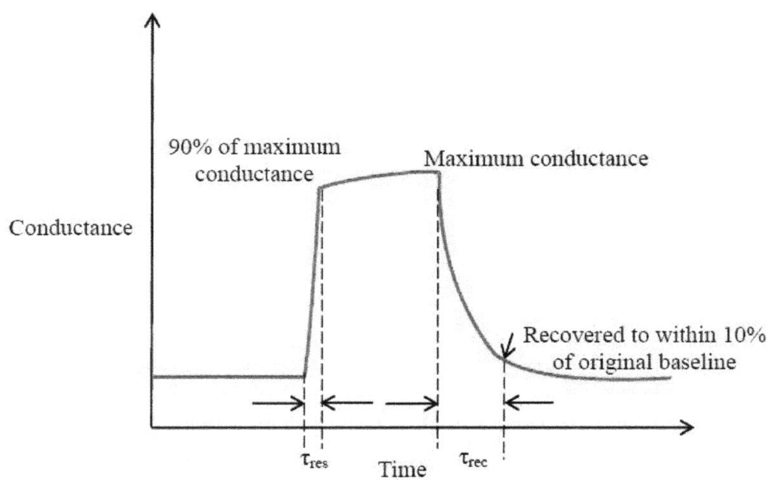

Figure 1.19 Recovery Time of Biosensors (Mushtaque Hussain, 2014)

Conclusion

Thus, sensors are becoming gradually more complicated, mostly due to a blend of advances in two technological fields like advanced technology & microelectronics. These are highly important devices to measure an extensive spectrum of analytes like gases, organic compounds, bacteria and ions. Thus, this is all about an overview of biosensors and the main components used in this sensor are physical components like amplifier and transducer whereas biological components like analyte and sensitive bio element. The characteristics of sensors mainly include linearity, sensitivity, selectivity, stability and response and recovery time.

References

Bhagwati Charan Patel, G R Sinha and Naveen Goel. Introduction to sensors, Advances in modern *sensors* 10 (2020) 7503-2707.

Wang Wen. Introductory Chapter: What is Chemical Sensors, 10 (2016) 5772 64626.

Roberto Paolesse, Sara Nardis, Donato Monti, Manuela Stefan Elli and. Prophyrinoids for Chemical sensing applications, *Chem. Rev 117* (2017) 2517-2583.

Hazal Gergeroglu, Serdar Yildirim, Mehmet Faruk Ebeoglugil. Nanocarbons in biosensor applications: an overview of carbon nanotubes (CNTs) and fullerenes (C_{60}), *SN Applied Sciences* (2020) 2:603.

Vikas Dull, Anjum Gahlut, Neeraj Dilbaghi. Acetylcholinesterase biosensors for electrochemical detection of Organophosphorus compounds, *Biochemistry Research International,* 1155 (2013) 731501.

Yan Zhou, Cheng Wei-Chou & Hong Liang. Interfacial Structures and Properties of Organic materials for Biosensors, *PubMed Central,* 11 (2012) 15036-15062.

Li Wang, David M.Sipe, MEMS thermal Biosensor for Metabolic monitoring Application, *Journal of Micro electrochemical system,* 10.1109 (2008) 916357.

Nishtha Khansili, Gurdeep Rattu, Prayaga Label-free optical biosensors for food and biological sensor applications, *Sensors & Actuators B: Chemicals,* 265 (2018) 35-49.

Chen Junsheng Wang. Optical Biosensors: An Exhaustive Overview, *Analyst* 145 (2020) 1605-1628.

Miroslav Pohanka. An Overview of Piezo-electric Biosensors, Immunosensors and DNA Sensors and their Application, *MDPI* 11 3 (2018) 448.

Miroslav Pohanka. Piezo-electric Biosensor for the determination of Tumour Necrosis factor Alpha, *Talanta* 178 (2018) 970-973.

Robert Magnusson, Debra Wawro, Yiwu Ding, Shelby Zimmerman. Resnant Photonic Biosensor with Polarization based multipara metric discrimination in each channel, Sensors (*Basel*) 11 2 (2001) 1476-88.

Briliant Adhi Prabowo, Paulo Freitas, Elisabete Fernandes The Challenge of developing Biosensors foe clinical application, *Chemosensors* 9 (2021) 299.

Bogdan Becure, Cristina Purcarea, Alina Vasileseu & Silvana Andreescu. Addressing the selectivity of Biosensors: Solutions & Perspectives, *Sensors,* 21 9 (2021) 3038.

M.Mohiuddin, D.Arbain, M.S.Ahmad & A.K.M. Shafiqul Islam. Alpha-Glucodase enzyme biosensor for the electrochemical measurement of Anti diabetic potential of medicinal plants, *National Research Letter* 11 1 (2016) 95.

Nizamettin Demirkiran, Ergun Ekinci & Meltem Asilturk. Immobilization of Glucose Oxidase in silica sol gel film for application of biosensor and Amperometric determination of Glucose, *J.Chi.Chem.Soc.* 4 (2012) 1336-1139.

Romain Guider, Davide Gandolfi, Alina Samusenko,. Sensitivity & Limit Detection of Biosensors based on ring resonators, *Sensing & Bio-sensing Research* 6 (2015) 99-102.

Mushtaque Hussain, Zafar Hussain Ibupoto, Omer Nur & Mazahr Ali Abbasi. Synthesis of three dimensional Nickel Cobalt Oxide Nano needles on nickel foam, their characterization and Glucose sensing application, *Sensors*, 14 3 (2014) 5415-25.

Chapter-2

ELECTROCHEMICAL BIOSENSOR

2.1 Overview of Electrochemistry

Electrochemistry is a branch of chemistry that studies the chemical reactions and changes that occur at the interface between an electrode and a solution. It is the study of the relationship between electricity and chemical reactions. The subject includes the study of electrochemical cells, which are devices that convert chemical energy into electrical energy or *vice versa*. Electrochemistry has many applications in various fields, including chemical synthesis, energy production, corrosion protection, electroplating, and analytical chemistry.

2.1.1 Applications of Electrochemistry

2.1.1.1 Batteries and Fuel Cells

One of the most significant applications of electrochemistry is in batteries and fuel cells, which are devices that store and convert energy as shown in Figure 2.11. Batteries convert chemical energy into electrical energy, while fuel cells convert chemical energy directly into electrical energy. Both types of devices consist of two electrodes, an anode, and a cathode, which are separated by an electrolyte. At the anode, a chemical reaction occurs, which generates electrons and ions. These electrons then flow through an external circuit to the cathode, where another chemical reaction occurs, which consumes the electrons and ions. The ions pass through the electrolyte to complete the circuit, and the overall reaction produces electrical energy.

2.1.1.2 Corrosion Protection

Electrochemistry is also used in corrosion protection, where a protective layer is applied to a metal surface to prevent corrosion. This process is called electroplating, and it involves the use of an electric current to deposit a thin layer of metal onto another metal surface.

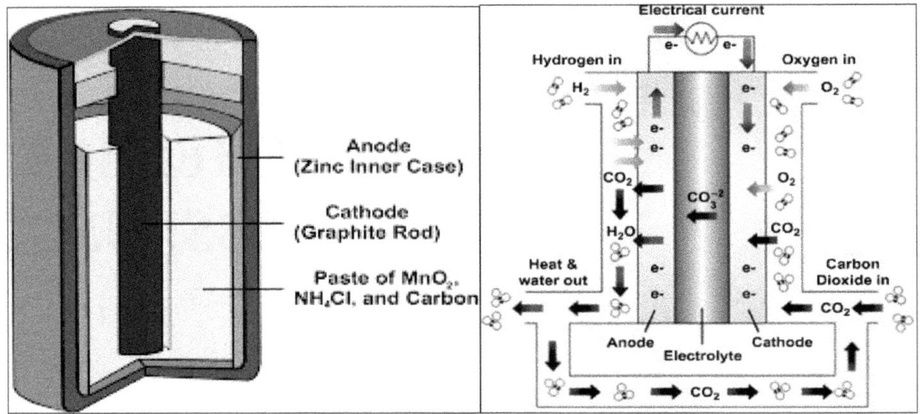

Figure 2.1 photograph image representation that (a) Batteries, (b) Fuel Cells (*From Web*)

2.1.1.3 Biosensors

In analytical chemistry, electrochemistry is widely used to detect and quantify chemical and biological species. Electrochemical techniques, such as cyclic voltammetry, amperometry, and impedance spectroscopy, are used in biosensors to detect and quantify analytes.

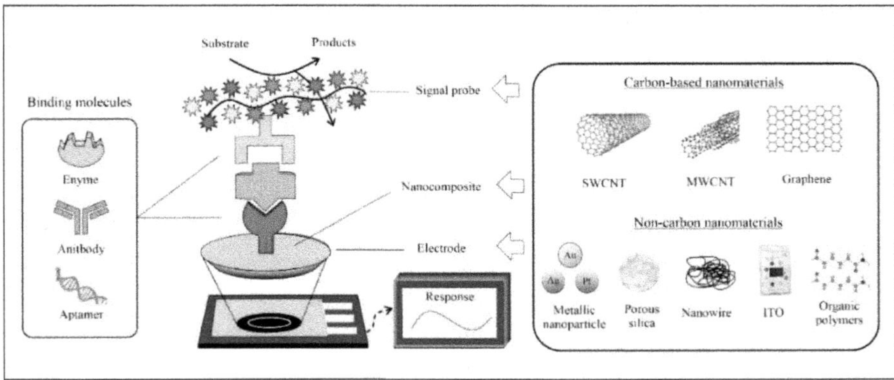

Figure 2.2 Scheme of analytical principle for electrochemical biosensors based on carbon and non-carbon nanomaterials

Biosensors are devices that use a biological recognition element, such as an enzyme and an antibody, to detect a specific analyte in a sample. The biological recognition element interacts with the analyte, producing a biochemical signal that is converted into an electrical signal by the electrochemical transducer as shown in Figure 2.2. The electrical signal is then processed to determine the concentration of the analyte in the sample. Electrochemical biosensors have many applications, including medical diagnostics, environmental monitoring, and food safety testing.

2.2 Electrochemical Biosensor

An electrochemical biosensor is a device that uses an electrochemical reaction to detect and measure the concentration of a particular substance, such as glucose, in a biological sample, as like blood or urine. The biological sensing element in an electrochemical biosensor can be an enzyme, antibody, nucleic acid, or whole cell, depending on the analyte being detected. The sensing element specifically interacts with the analyte, producing a biochemical signal that is proportional to its concentration in the sample. The electrochemical transducer is responsible for converting the biochemical signal into an electrical signal, which can be measured and quantified. The most common electrochemical transducers used in biosensors are voltammetry, amperometric, and potentiometric, etc., Electrochemical biosensors have gained significant attention in recent years due to their high sensitivity, low cost, and portability. They have a wide range of applications in fields such as clinical diagnostics, environmental monitoring, food safety testing, and biosecurity.

2.3 Principles of Electrochemical Biosensors

The principles of electrochemical biosensors involve the integration of a biological recognition element with an electrochemical transducer to produce a measurable electrical signal that is proportional to the concentration of a target analyte in a sample. The basic components of an electrochemical biosensor include:

- **Biological recognition element**: This is a biological molecule or entity that specifically binds to the target analyte. It can be an enzyme, antibody, nucleic acid, or whole cell, depending on the target analyte.
- **Transducer**: This is a device that converts the biochemical signal generated by the biological recognition element into an electrical signal. The most common types of transducers used in electrochemical biosensors are amperometric, potentiometric, and impedimetric.
- **Signal processing system**: This is a device that analyses and quantifies the electrical signal produced by the transducer to determine the concentration of the target analyte in the sample.

The operation of an electrochemical biosensor involves the interaction between the biological recognition element and the target analyte as shown in Figure 2.3. This interaction generates a biochemical signal, such as a change in pH, current, or voltage. The transducer converts this signal into an electrical signal that can be measured and quantified. The signal processing system then uses this electrical signal to determine the concentration of the target analyte in the sample.

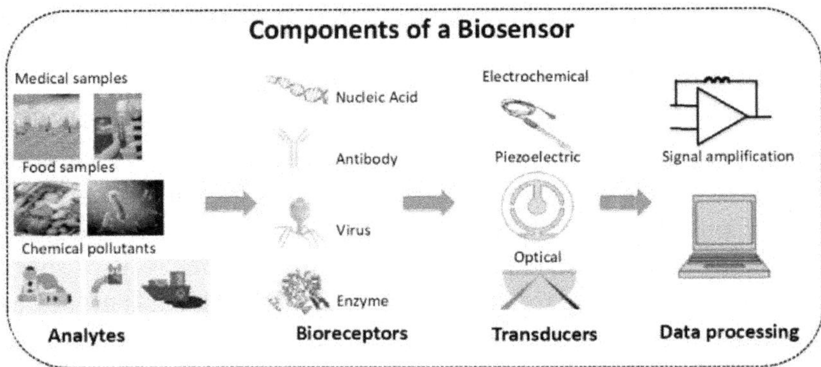

Figure 2.3 Schematic representation that the basic principle of electrochemical biosensor

The sensitivity and selectivity of an electrochemical biosensor depend on the specificity and affinity of the biological recognition element for the target analyte, as well as the performance characteristics of the transducer and signal processing system. Electrochemical biosensors have a wide range of applications in various fields, such as medical diagnostics, environmental monitoring, and food safety testing, among others.

Cyclic voltammetry

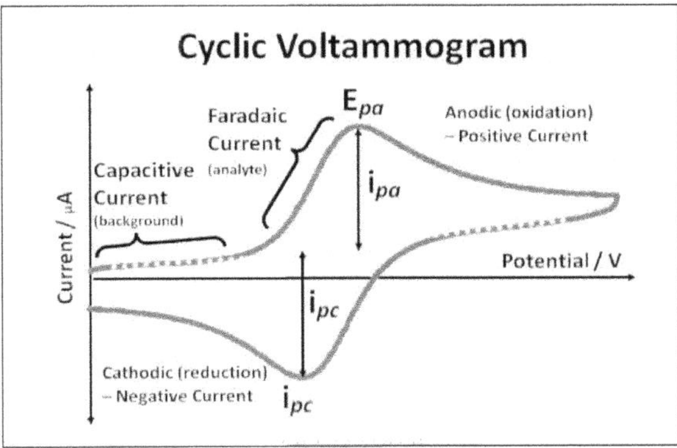

Figure 2.4 Schematic representation that the cyclic voltammogram

Cyclic voltammetry is an electrochemical technique that is widely used in analytical chemistry. It involves the application of a potential to an electrode, and the measurement of the resulting current as the potential is scanned as shown in Figure 2.4. The technique is used to study the redox behaviour of a system and to determine the kinetic parameters of electrochemical reactions.

Amperometry

Amperometry is another electrochemical technique that is widely used in analytical chemistry. It involves the measurement of current at a fixed potential, and it is used to detect and quantify analytes in a sample. In amperometry, a constant potential is applied to the electrode, and the current is measured as the analyte is oxidized or reduced at the electrode. Amperometry involves the measurements of currents at constant voltage applied at the dropping mercury electrode as shown in Figure 2.5. The value of electrode potential is chosen in such a way that only the metal ion is reduced. This method is generally used for the determination of metal ion present in aqueous solution.

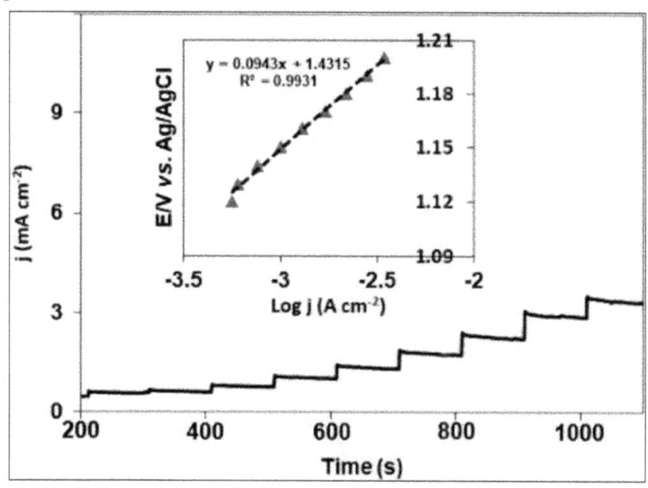

Figure 2.5 Schematic representation that the amperometry

Impedance spectroscopy

Impedance spectroscopy is an electrochemical technique that is used to measure the resistance of a system to an applied electrical current as shown in Figure 2.6. The technique is used to study the electrochemical properties of a system and to determine the kinetic parameters of electrochemical reactions. Impedance spectroscopy is widely used in biosensors and other analytical instruments. Overall, electrochemistry plays a vital role in many areas of science and technology, and its principles are essential for understanding the operation of electrochemical biosensors and other electrochemical devices. Electrochemical techniques are widely used in analytical chemistry, and they have many applications in fields such as energy production, corrosion protection, and materials science. As research continues, new and exciting applications of electrochemistry are likely to be discovered.

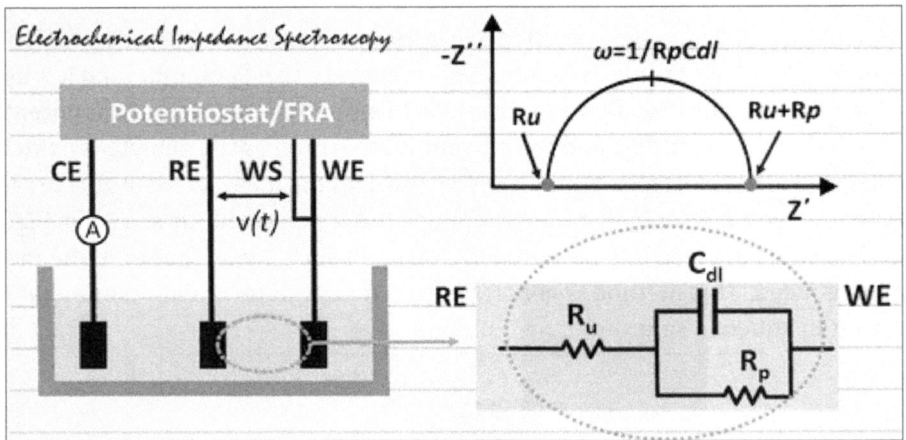

Figure 2.6 Schematic representation that the Impedance spectroscopy

2.4 Conductometry

Conductometry is a branch of analytical chemistry that deals with the measurement of electrical conductivity. Electrical conductivity is the ability of a solution to conduct an electrical current, and it is related to the concentration and mobility of ions in the solution. Conductometry is widely used in analytical chemistry to determine the concentration of ions and molecules in a sample. In conductometry, an electrical potential is applied to a solution containing ions, and the resulting electrical current is measured.

The electrical conductivity of the solution is directly proportional to the concentration of ions in the solution. Conductometry is a sensitive technique, and it can detect very low concentrations of ions in a solution.4 Conductivity measurements can be carried out using a variety of instruments, including conductivity meters and electrodes. Conductivity meters are devices that measure the electrical conductivity of a solution, while electrodes are devices that are used to apply an electrical potential to a solution.

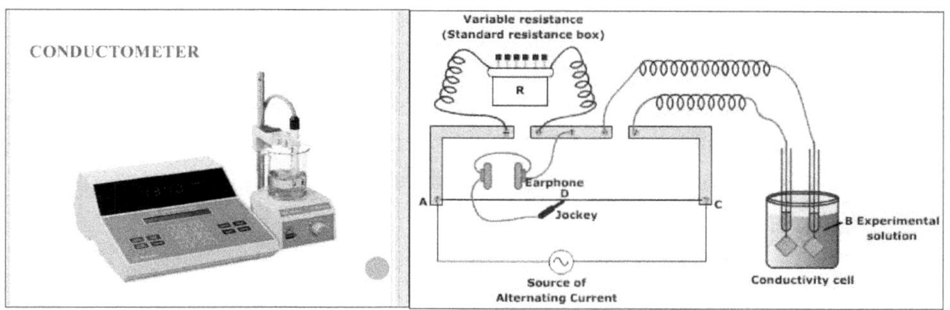

Figure 2.7 Photograph image of conductometry instrument

2.4.1 Principle

The principles of conductometry are based on Ohm's law, (V=IR) which states that the current flowing through a conductor is directly proportional to the voltage applied across the conductor. The electrical conductivity of a solution can be calculated by measuring the current flowing through the solution and the voltage applied across the solution. The conductivity of the solution is given by the ratio of the current to the voltage.

2.4.2 Applications

Conductometry is used in a wide range of applications, including environmental monitoring, quality control, and pharmaceutical analysis. It is commonly used to determine the concentration of ions in a sample, such as the concentration of chloride ions in water. Conductivity measurements can also be used to study the kinetics of chemical reactions and to determine the strength of acids and bases. One important application of conductometry is in the pharmaceutical industry, where it is used to determine the purity of drugs and to monitor the progress of chemical reactions. Conductivity measurements can also be used to study the interactions between drugs and biomolecules, such as proteins and DNA.

Another application of conductometry is in environmental monitoring, where it is used to measure the conductivity of water and soil samples. Conductivity measurements can be used to detect the presence of ions and other contaminants in the environment. In summary, conductometry is a useful technique in analytical chemistry that is used to measure the electrical conductivity of a solution. The technique is based on Ohm's law, and it is widely used to determine the concentration of ions and molecules in a sample. Conductivity measurements have many applications in environmental monitoring, quality control, and pharmaceutical analysis.

2.5 Voltammetry

Voltammetry is a type of electrochemical analysis that measures the current as a function of the applied potential. It is a powerful analytical technique that is widely used in chemistry, biochemistry, and material science to study redox reactions, electrochemical behavior of molecules, and kinetics of chemical reactions.

2.5.1 Principle

The basic principle of voltammetry is to apply a voltage or potential to a working electrode immersed in a solution containing the sample of interest, and measure the resulting current flowing through the electrode. The potential is swept or stepped over a range of values, and the current is recorded at each potential. The resulting current-potential curve is known as a voltammogram, which provides information about the electrochemical properties of the sample as shown in Figure 2.8.

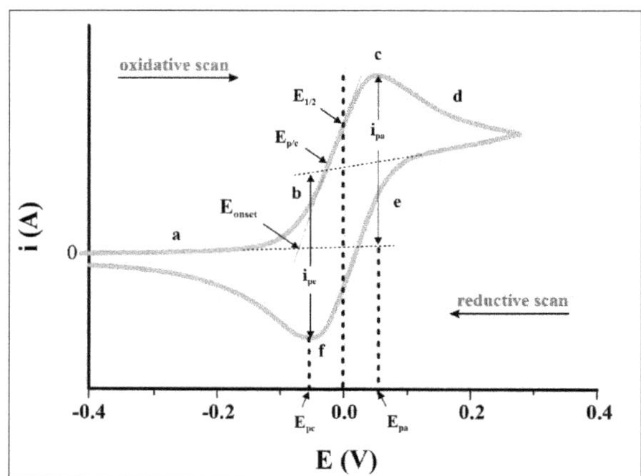

Figure 2.8 Cyclic voltammogram for an electrochemically reversible one-electron redox process

A cyclic voltammogram is the 'duck-shaped' plot generated by cyclic voltammetry. In the example cyclic voltammogram above, the scan starts at -0.4V and sweeps forward to more positive, oxidative potentials. Initially the potential is not sufficient to oxidise the analyte (a). As the potential approaches several kT of the standard potential, the onset (E_{onset}) of oxidation is reached. Following this, the current exponentially increases (b) as the analyte begins its oxidation at the working electrode surface. For a reversible process, here the current rises initially as if there is no change in the concentration of oxidant. The current is dictated by the rate of diffusion of the oxidant to the electrode, as well as the proportion converted to the reduced form. This can be understood according to the Nernst equation. Gradually, as the scan continues, more oxidant is depleted. The concentration gradient adjusts to this. It is this change which causes a peak in the voltammogram. You can see how the decrease in current from depletion of the oxidant outweighs the increase from changing the proportion of oxidant oxidised at the electrode. The current reaches peak maximum at point c (anodic peak current (i_{pa}) for oxidation at the anodic peak potential (E_{pa}). Here, more positive potentials cause an increase in current that is offset by a decreasing flux of analyte from further and further distance from the electrode surface.

From this point the current is limited by the mass transport of analyte from the bulk to the DDL interface, which is slow on the electrochemical timescale. This results in a decrease in current (d), as the potentials are scanned more positive. This occurs until a steady-state is reached where further increases in potential no longer has an effect. Scan reversal to negative potentials (reductive scan) continues to oxidise the analyte. This continues until the applied potential reaches the value where the oxidised analyte (which has accumulated at the electrode surface) can be re-reduced (e). The process for reduction mirrors that for the oxidation. The only difference is that it occurs with an opposite scan direction and a cathodic peak (i_{pc}) at the cathodic peak potential (E_{pc}) (f). The anodic and cathodic peak currents should be of equal magnitude but with opposite sign. This is only provided that the process is reversible (and if the cathodic peak is measured relative to the base line after the anodic peak).

2.5.2 Types of Voltammetry

Voltammetry can be classified into various types based on the nature of the working electrode and the type of measurement. The most common types of voltammetry include cyclic voltammetry, differential pulse voltammetry, square wave voltammetry, and polarography.

2.5.2.1 Cyclic voltammetry

Cyclic voltammetry (CV) is a technique where the potential is swept back and forth between two values, and the resulting current is measured. CV is useful for studying redox reactions, electrochemical kinetics, and surface electrochemistry. The shape of the resulting CV curve provides information about the thermodynamics and kinetics of the reaction.

2.5.2.2 Differential Pulse Voltammetry

Differential pulse voltammetry (DPV) is a technique where a small pulse of potential is applied to the working electrode, and the resulting current is measured. DPV is useful for detecting trace amounts of analytes and for studying fast reactions.

2.5.2.3 Square Wave Voltammetry

Square wave voltammetry (SWV) is a technique where a square wave of potential is applied to the working electrode, and the resulting current is measured. SWV is useful for studying complex reactions and for detecting low concentrations of analytes.

2.5.2.4 Polarography

Polarography is a type of voltammetry where the potential is swept linearly over a range of values, and the resulting current is measured. Polarography is useful for studying the reduction and oxidation of ions and for determining the concentration of dissolved gases.

2.5.3 Applications

Voltammetry has many applications in chemistry, biochemistry, and material science. It is commonly used to study the electrochemical behavior of molecules, such as proteins, DNA, and drugs. Voltammetry is also used in the development of new materials for energy storage and in the design of electrochemical sensors for environmental monitoring and medical diagnostics. In summary, voltammetry is a powerful analytical technique used to study redox reactions and electrochemical behavior of molecules. It involves applying a voltage to a working electrode and measuring the resulting current. Voltammetry has many applications in chemistry, biochemistry, and material science, and it is a valuable tool for the development of new materials and the design of electrochemical sensors.

2.6 Impedance Bio-sensors

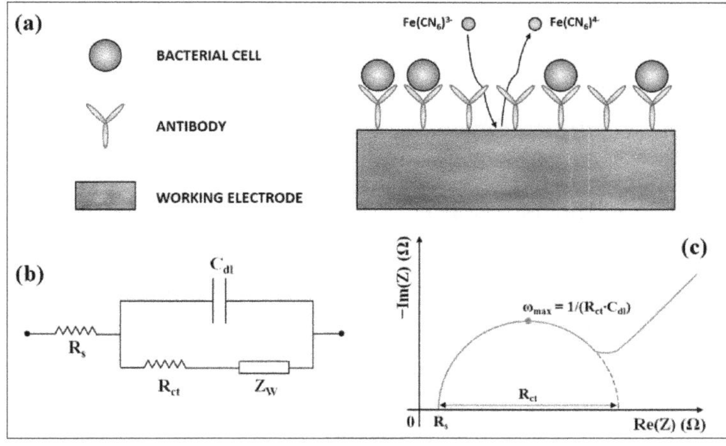

Figure 2.9 Schematic illustration representation that the working principle of impedance

Impedance biosensors are a type of biosensor that work on the principle of measuring the impedance or resistance of an electrical circuit that includes a biological element. These sensors are widely used in biomedical research, clinical diagnostics, and environmental monitoring as shown in Figure 2.9.

2.6.1 Principle

The basic principle of an impedance biosensor involves the immobilization of a biological element, such as enzymes, antibodies, or DNA, on an electrode surface. When the biological element interacts with a target analyte, it leads to a change in the electrical properties of the electrode, such as impedance or capacitance. This change in the electrical properties is measured and used to detect the presence and concentration of the analyte.

2.6.2 Types of Impedance biosensors

Impedance biosensors can be classified into two types based on the measurement technique used: frequency domain and time domain, In frequency domain measurements, a small alternating current is applied to the electrode at a fixed frequency, and the resulting impedance is measured. The change in impedance due to the interaction between the biological element and the analyte is measured and used to determine the concentration of the analyte. In time domain measurements, a small voltage pulse is applied to the electrode, and the resulting current is measured over a fixed period of time. The change in the current due to the interaction between the biological element and the analyte is measured and used to determine the concentration of the analyte.

2.6.3 Applications

Impedance biosensors have several advantages over other types of biosensors. They are label-free, meaning that no additional labeling or tagging of the analyte is required. They are also highly sensitive and can detect very low concentrations of analytes, making them useful in a wide range of applications. One of the key applications of impedance biosensors is in medical diagnostics. They are used to detect biomarkers of diseases, such as cancer and infectious diseases, and to monitor the progression of these diseases. Impedance biosensors can also be used to detect toxins and pathogens in food and water samples, making them useful in environmental monitoring. In summary, impedance biosensors are a type of biosensor that measure the electrical impedance or resistance of an electrical circuit that includes a biological element. They are highly sensitive and can detect very low concentrations of analytes, making them useful in a wide range of applications, including medical diagnostics and environmental monitoring.

Impedance biosensors are label-free and have several advantages over other types of biosensors, making them an attractive option for researchers and practitioners.

2.7 Semiconductors

Semiconductors are materials that have electrical conductivity between that of conductors, such as metals, and insulators, such as ceramics or plastics. They are an important class of materials that are widely used in electronic devices, such as computer chips, solar cells, and light-emitting diodes (LEDs). The electrical properties of semiconductors are determined by their band structure as shown in Figure 2.10. In a semiconductor, the valence band, which is filled with electrons, is separated from the conduction band, which is empty or partially filled, by a bandgap. The energy required to excite an electron from the valence band to the conduction band is equal to the bandgap energy.

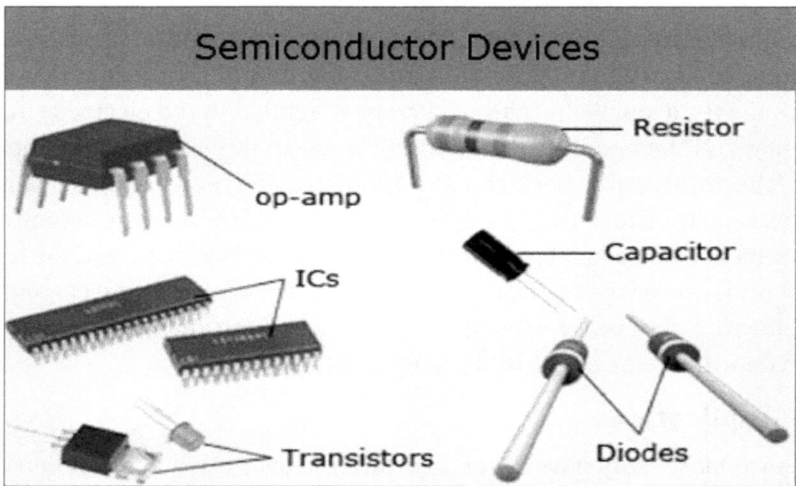

Figure 2.10 Photograph image representation that the Semiconductor components

2.7.1 Types of Semiconductors

Semiconductors can be classified into two types: intrinsic and extrinsic.

Intrinsic semiconductors, such as silicon and germanium, are pure materials that have a low concentration of impurities. The electrical conductivity of intrinsic semiconductors can be increased by adding impurities, a process known as doping. Extrinsic semiconductors can be further classified into two types: n-type and p-type as shown in Figure 2.11. In n-type semiconductors, such as silicon doped with arsenic, the dopant atoms have extra electrons that are free to move in the conduction band. In p-type semiconductors, such as silicon doped with boron, the dopant atoms have fewer electrons than the host atoms, creating holes in the valence band that act as positive charge carriers.

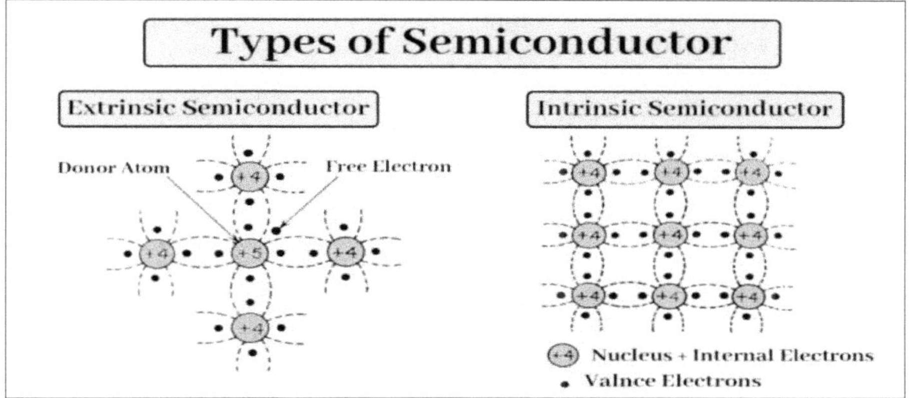

Figure 2.11 Types of Semiconductors (Palash Moni Saikia, May 31, 2022)

2.7.2 Applications

Semiconductors are used in a wide range of electronic devices, including diodes, transistors, and integrated circuits. In a diode, a p-n junction is formed by joining a p-type semiconductor to an n-type semiconductor. When a voltage is applied across the diode, electrons flow from the n-type side to the p-type side, creating a current. In a transistor, a small voltage applied to the base of the device controls the flow of current through the device. Semiconductors are also used in solar cells, which convert sunlight into electrical energy. Solar cells are typically made of silicon or other semiconductor materials, and work by absorbing photons of light, which excite electrons and create a flow of current. In addition to electronic devices, semiconductors are used in optoelectronic devices, such as LEDs and photodiodes. LEDs emit light when a current is passed through them, while photodiodes detect light and convert it into a current. In summary, semiconductors are materials that have electrical conductivity between that of conductors and insulators. They are widely used in electronic devices, such as diodes, transistors, and solar cells, as well as optoelectronic devices, such as LEDs and photodiodes. The electrical properties of semiconductors are determined by their band structure, and can be controlled by doping with impurities.

2.8 Ion-Selective Field-Effect Transistor (ISFET)

An ion-selective field-effect transistor (ISFET) is a type of transistor that is sensitive to the presence of ions in a solution. It is a semiconductor device that is used to measure the concentration of ions in a liquid or gas. Source and drain are the two electrodes used in a FET system. The electron flow takes place in a channel between the drain and source. The gate potential controls the flow of current between the two electrodes.

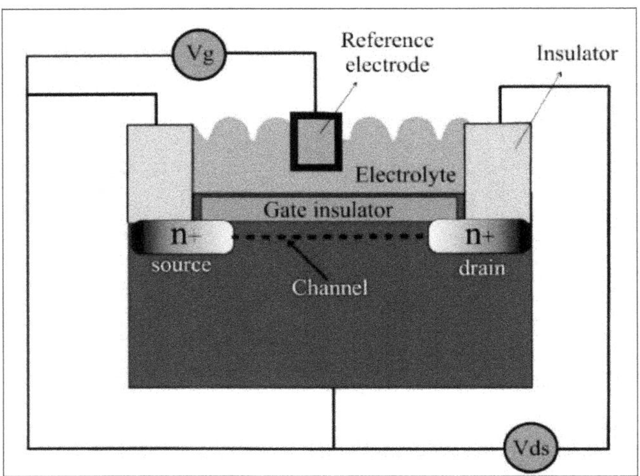

Figure 2.12 Schematic view of an Ion-Selective Field-Effect Transistor

2.8.1 Principle

The ISFET is based on the same basic principle as a metal-oxide-semiconductor field-effect transistor (MOSFET), which is a type of transistor used in many electronic devices. The difference is that the gate of an ISFET is covered with a layer of an ion-sensitive material, such as a polymer or oxide, which responds to changes in the concentration of ions in the solution.

When the ISFET is placed in a solution containing ions, a potential difference is created between the ion-sensitive layer and the solution. This potential difference changes the conductivity of the semiconductor material beneath the ion-sensitive layer, which is detected as a change in the current flowing through the transistor. The response of an ISFET to a particular ion depends on the ion-sensitive material used. For example, a layer of silicon dioxide can be used to detect hydrogen ions (pH), while a layer of titanium dioxide can be used to detect potassium ions (K^+).

2.8.2 Applications

ISFETs are widely used in chemical and biological sensors, as they are highly sensitive and can be used to detect a wide range of ions. They are also small and easy to integrate into electronic circuits, making them ideal for use in portable devices.

2.8.3 Limitation of ISFETs

One limitation of ISFETs is that they are sensitive to changes in temperature, which can affect the accuracy of the measurements. To overcome this, temperature compensation circuits can be used to adjust the measurements based on the temperature of the solution.

In summary, an ion-selective field-effect transistor (ISFET) is a semiconductor device that is sensitive to the presence of ions in a solution. It works by detecting changes in the conductivity of the semiconductor material beneath an ion-sensitive layer on the gate of the transistor. ISFETs are widely used in chemical and biological sensors, and are highly sensitive and small enough to be integrated into electronic circuits. However, they are sensitive to changes in temperature, which can affect their accuracy.

2.9 Enzyme Field Effect Transistor (EnFET)

An enzyme field effect transistor (EnFET) is a biosensor device that uses an enzyme as a bioreceptor to detect and quantify specific analytes. It is a type of field effect transistor (FET) that measures changes in the electrical properties of the device in response to the binding of the analyte to the enzyme.

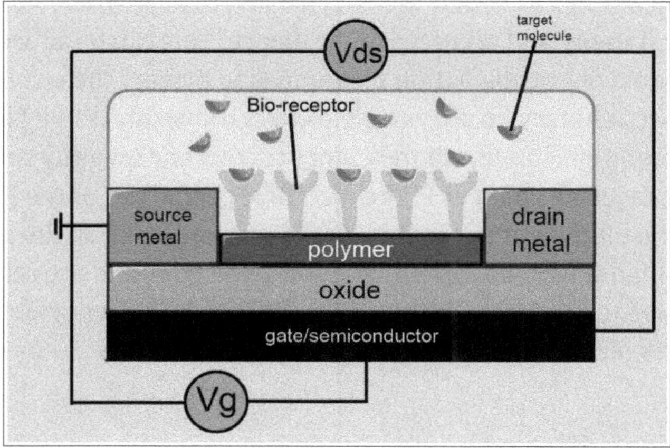

Figure 2.13 The schematic view of an Enzyme Field Effect Transistor

In a typical BioFET, an electrically and chemically insulating layer (e.g. Silica) separates the analyte solution from the semiconducting device. A polymer layer, most commonly APTES, is used to chemically link the surface to a receptor which is specific to the analyte (e.g. biotin or an antibody). Upon binding of the analyte, changes in the electrostatic potential at the surface of the electrolyte-insulator layer occur, which in turn results in an electrostatic gating effect of the semiconductor device, and a measurable change in current between the source and drain electrodes.

2.9.1 Working

The EnFET consists of a gate electrode coated with an enzyme layer that is specific to the analyte of interest. When the analyte binds to the enzyme, a change in the electrical charge or potential is produced, which modulates the conductivity of the FET channel. This change in conductivity can be measured and correlated to the concentration of the analyte in the sample.

2.9.2 Applications

EnFETs have several advantages over other biosensor technologies, such as their high sensitivity, specificity, and selectivity. They are also relatively simple and inexpensive to fabricate, and can be easily integrated into electronic circuits. EnFETs can be used to detect a wide range of analytes, such as glucose, lactate, cholesterol, and neurotransmitters. They have applications in clinical diagnostics, environmental monitoring, and food safety testing.

2.9.3 Limitation of EnFETs

One limitation of EnFETs is that the enzyme layer can degrade over time, which can affect the stability and accuracy of the device. This can be addressed by using stabilizing agents or encapsulation techniques to protect the enzyme layer from degradation. In summary, an enzyme field effect transistor (EnFET) is a biosensor device that uses an enzyme as a bioreceptor to detect and quantify specific analytes. It works by measuring changes in the electrical properties of the FET channel in response to the binding of the analyte to the enzyme layer on the gate electrode. EnFETs have high sensitivity, specificity, and selectivity, and are relatively simple and inexpensive to fabricate. However, the stability of the enzyme layer can be a limitation, which can be addressed by using stabilizing agents or encapsulation techniques.

2.10 Glucose Biosensor

A glucose biosensor is a device that is used to measure the concentration of glucose in a sample. It is a type of biosensor that uses an enzyme as a bioreceptor to specifically detect glucose.

2.10.1 Working Mechanism

Glucose biosensors work by using the enzyme glucose oxidase (GOx), which reacts with glucose to produce gluconic acid and hydrogen peroxide. The hydrogen peroxide is then detected using an electrochemical or optical transducer.

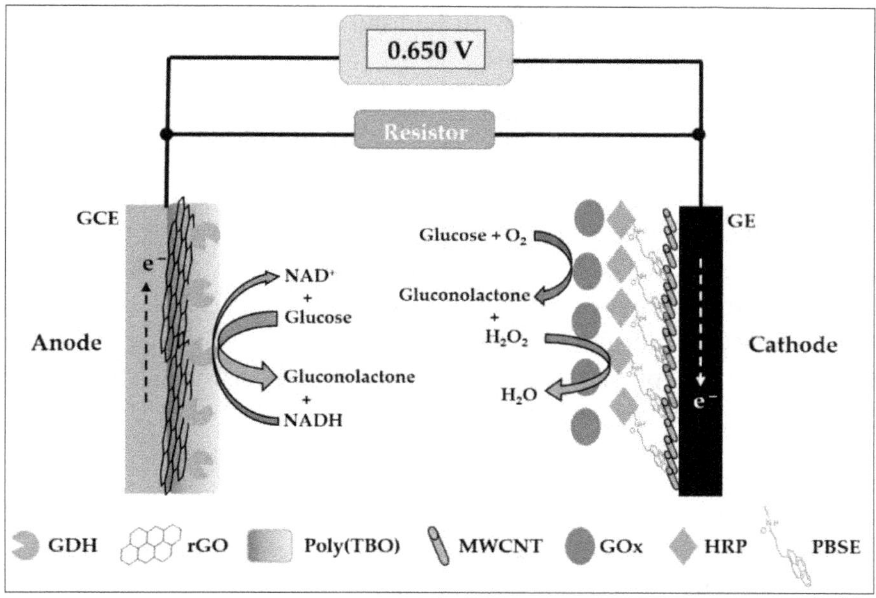

Figure 2.14 Scheme of the glucose self-powered biosensor developed in this work based on an enzymatic biofuel cell.

2.10.1.1 Electrochemical Glucose Biosensor

The electrochemical glucose biosensor is the most common type of glucose biosensor. It consists of a glucose-sensitive electrode that is coated with a layer of glucose oxidase. When glucose is present, it reacts with the glucose oxidase to produce hydrogen peroxide, which is then detected by measuring the current or voltage generated by the electrochemical reaction.

2.10.1.2 Optical Glucose Biosensors

Optical glucose biosensors work by using a fluorescence or colorimetric method to detect the hydrogen peroxide produced by the reaction between glucose and glucose oxidase. In a fluorescence biosensor, a fluorescent molecule is used to detect the hydrogen peroxide, while in a colorimetric biosensor, a dye is used to produce a color change in response to the hydrogen peroxide.

2.10.2 Applications

Glucose biosensors have a wide range of applications in the medical field, particularly in the management of diabetes. Glucose biosensors can be used to monitor blood glucose levels in diabetic patients, allowing for better control of the disease and reducing the risk of complications.In addition to medical applications, glucose biosensors have applications in food and beverage analysis, environmental monitoring, and biofuel production.

2.10.3 Limitation of Glucose Biosensors

One limitation of glucose biosensors is that interferents, such as other sugars, which can produce false readings, can affect them. To overcome this limitation, glucose biosensors can be designed to use multiple enzymes or selective membranes to improve their selectivity. In summary, glucose biosensors are devices that are used to measure the concentration of glucose in a sample. They use the enzyme glucose oxidase to specifically detect glucose and can work through either an electrochemical or optical transducer. Glucose biosensors have a wide range of applications in the medical field and beyond, but are limited by their selectivity in the presence of interferents.

2.11 Nanostructure Materials in Electrochemical Biosensors

Nanostructure materials are increasingly being used in the development of electrochemical biosensors due to their unique properties, such as high surface area, enhanced conductivity, and increased sensitivity as shown in figure 2.15. Some of the commonly used nanostructure materials in electrochemical biosensors are:

- **Carbon-based nanostructures:** Carbon nanotubes (CNTs) and graphene are widely used in electrochemical biosensors due to their excellent electrical conductivity and large surface area. CNTs can act as a nanowire to increase electron transfer rate and provide a high surface area for immobilizing bioreceptors. Graphene, on the other hand, has high sensitivity and selectivity due to its high surface area and the ability to form strong interactions with biological molecules.
- **Metal nanoparticles:** Metal nanoparticles such as gold, silver, and platinum are widely used as electrode materials in electrochemical biosensors due to their excellent catalytic properties and high surface area. These nanoparticles can be used to enhance electron transfer rate, increase sensitivity, and improve the stability of the electrode.
- **Metal oxide nanostructures:** Metal oxide nanostructures such as titanium dioxide (TiO_2), zinc oxide (ZnO), and iron oxide (Fe_2O_3) are also used in electrochemical biosensors due to their unique properties, such as high surface area, tunable electronic properties, and excellent catalytic activity. These nanostructures can act as a matrix for immobilizing bioreceptors and enhance the sensitivity of the biosensor.

- **Quantum dots:** Quantum dots (QDs) are semiconductor nanoparticles that have unique optical and electronic properties. They are used in electrochemical biosensors as labels for signal amplification and detection. QDs can be conjugated with bioreceptors to increase the sensitivity of the biosensor.
- **Molecularly imprinted polymers:** Molecularly imprinted polymers (MIPs) are synthetic materials that can selectively bind to target molecules. They are used in electrochemical biosensors as recognition elements due to their high selectivity and sensitivity. MIPs can be designed to have a high affinity for specific molecules, making them useful for detecting a wide range of analytes.

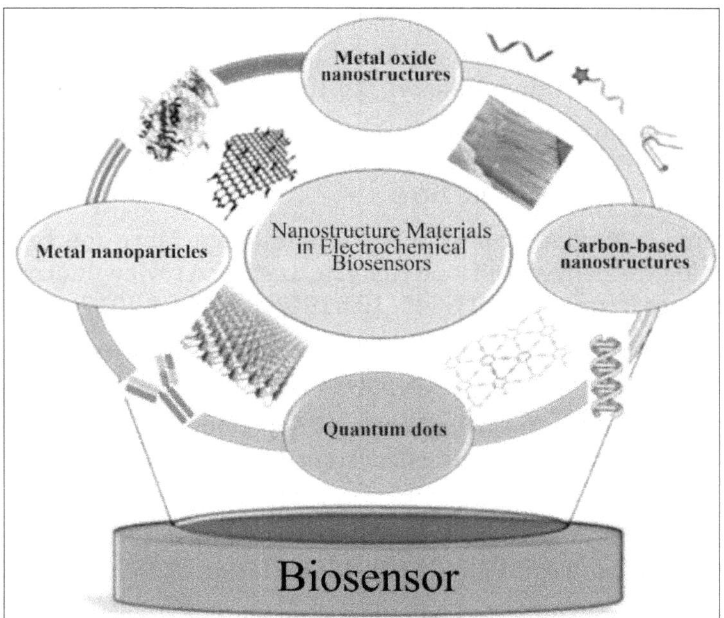

Figure 2.15 Nanostructure materials in electrochemical biosensors

In summary, nanostructure materials are being increasingly used in electrochemical biosensors due to their unique properties. Carbon-based nanostructures, metal nanoparticles, metal oxide nanostructures, quantum dots, and molecularly imprinted polymers are some of the commonly used nanostructure materials in electrochemical biosensors. These materials can enhance the sensitivity, selectivity, and stability of electrochemical biosensors and have a wide range of applications in various fields, including clinical diagnostics, environmental monitoring, and food safety testing.

References

Li, Haitao, et al. "CMOS electrochemical instrumentation for biosensor microsystems: A review." Sensors 17.1 (2016): 74.

Brett Christopher M. A and Ana Maria Oliveira Brett. Electrochemistry: Principles Methods and Applications. Oxford University Press 1993.

Badwal, Sukhvinder P S et al. "Emerging electrochemical energy conversion and storage technologies." Frontiers in chemistry vol. 2, 24 Sep. 2014, doi:10.3389/fchem.2014.00079

Khopkar, S.M., "Basic Concepts of Analytical Chemistry", 3rd edition, 2007, ISBN 978-81-224-2092-0.

Serban C. Moldoveanu, Victor David, Chapter 2 - Short Overviews of Analytical Techniques Not Containing an Independent Separation Step, Selection of the HPLC Method in Chemical Analysis, Elsevier, 2017, Pages 31-53, ISBN 9780128036846, https://doi.org/10.1016/B978-0-12-803684-6.00002-0.

N. Elgrishi, K. J. Rountree, B. D. McCarthy, E. S. Rountree, T. T. Eisenhart, and J. L. Dempsey *A Practical Beginner's Guide to Cyclic Voltammetry* J. Chem. Educ., vol. 95, no. 2, pp. 197–206, 2018

Grossi, Marco & Riccò, Bruno. (2017). Electrical impedance spectroscopy (EIS) for biological analysis and food characterization: A review. Journal of Sensors and Sensor Systems. 6. 303-325. 10.5194/jsss-6-303-2017.

Shockley, William (1950). Electrons and holes in semiconductors: with applications to transistor electronics. R. E. Krieger Pub. Co. ISBN 978-0-88275-382-9.

Schöning, Michael J.; Poghossian, Arshak (10 September 2002). "Recent advances in biologically sensitive field-effect transistors (BioFETs)". Analyst. 127 (9): 1137–1151. doi:10.1039/B204444G. ISSN 1364-5528. PMID 12375833.

Yoo, Eun-Hyung, Soo-Youn Lee. "Glucose biosensors: an overview of use in clinical practice." Sensors vol. 10,5 (2010): 4558-76. doi:10.3390/s100504558

Chansaenpak, K.; Kamkaew, A.; Lisnund, S.; Prachai, P.; Ratwirunkit, P.; Jingpho, T.; Blay, V.; Development of a Sensitive Self-Powered Glucose Biosensor Based on an Enzymatic Biofuel Cell. Biosensors 2021, 11, 16. https://doi.org/10.3390/bios11010016

Nekane Reta, Christopher P. Saint, Andrew Michelmore Orcid, Beatriz Prieto-Simon, and Nicolas H. Voelcker Orcid, "Nanostructured Electrochemical Biosensors for Label-Free Detection of Water- and Food-Borne Pathogens", CS Appl. Mater. Interfaces 2018, 10, 7, 6055–6072, Publication Date: January 25, 2018, https://doi.org/10.1021/acsami.7b13943

Chapter-3

OPTICAL AND COLORIMETRIC BIOSENSORS

3.1 Introduction

Optical and colorimetric biosensors are another type of biosensors that are commonly used in the field of biotechnology and medical research. Optical biosensors are devices that use light to detect and analyse biological molecules or analytes. These devices typically consist of a biological recognition element, such as an antibody or enzyme, that is immobilized onto a solid surface, and a signal transduction element, such as a fluorescent or luminescent dye, that responds to the presence of the analyte by changing its optical properties. The changes in the optical properties can be measured and correlated to the concentration of the analyte.

Colorimetric biosensors, on the other hand, are a type of optical biosensor that rely on changes in colour to detect and quantify analytes. These devices typically use a colorimetric indicator, such as a dye or nanoparticle, that changes colour in the presence of the analyte. The colour change is then measured and correlated to the concentration of the analyte. Both optical and colorimetric biosensors have advantages and disadvantages, and the choice of which to use depends on the specific application. Optical biosensors are generally more sensitive and specific, but they require more complex instrumentation and are often more expensive. Colorimetric biosensors, on the other hand, are relatively simple to use and require less instrumentation, but they may not be as sensitive or specific as optical biosensors.

3.2 Principles of Optical Biosensors

Optical biosensors work on the principle of detecting changes in the optical properties of a material in response to the presence of a target analyte as shown Figure 3.1. The basic components of an optical biosensor are:

- **Biological recognition element:** This is a biologically active molecule, such as an enzyme, antibody, or nucleic acid, that binds specifically to the target analyte.
- **Transducer:** This is a material that converts the binding event between the biological recognition element and the analyte into a measurable signal. Common transducers include fluorophores, metal nanoparticles, and surface plasmon resonance (SPR) sensors.
- **Optical Detection System:** This is a device that detects the signal from the transducer and provides a readout of the concentration of the target analyte.

There are several types of optical biosensors, each with its unique mechanism of action. Some common examples are:

- **Fluorescence-based Biosensors:** These biosensors use a fluorophore, which emits light at a specific wavelength when excited by a light source. Binding of the analyte to the biological recognition element causes a change in the fluorescent signal, which can be detected and quantified.
- **Surface Plasmon Resonance (SPR)-based Biosensors:** These biosensors use a metal surface to detect binding events between the biological recognition element and the analyte. Binding causes a change in the refractive index of the metal surface, which can be detected as a shift in the SPR signal.
- **Optical Fiber-based Biosensors:** These biosensors use optical fibers coated with the biological recognition element to detect the target analyte. Binding of the analyte causes a change in the refractive index of the coating, which can be detected as a change in the light transmitted through the fiber.

Overall, optical biosensors offer high sensitivity and specificity, rapid response times, and the ability to detect a wide range of analytes. They have a wide range of applications, including medical diagnostics, environmental monitoring, and food safety testing.

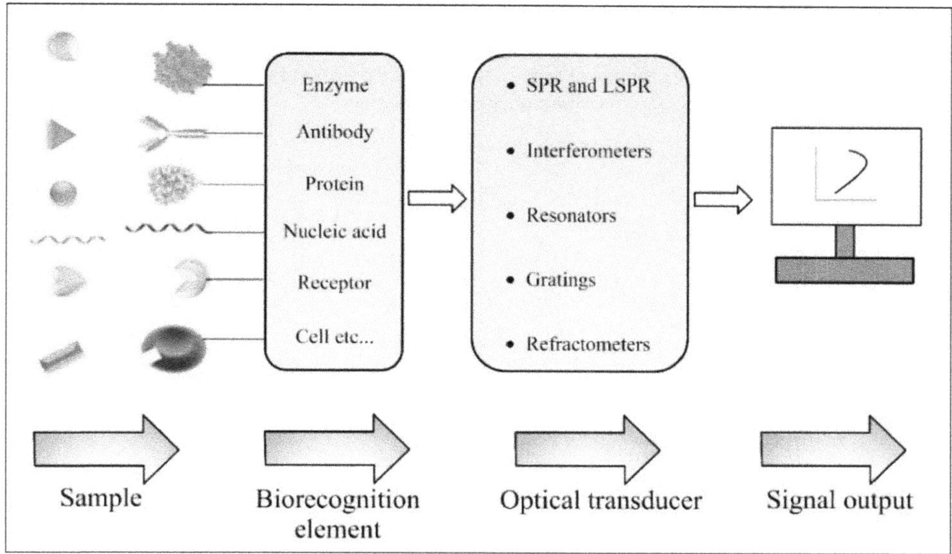

Figure 3.1 Schematic illustration representation that the basic principle of optical biosensor

3.3 Absorbance

Absorbance is a measure of the amount of light that is absorbed by a material at a specific wavelength. It is defined as the logarithm of the ratio of the incident light intensity to the transmitted light intensity, and is expressed in units of optical density or absorbance units (AU). The absorbance of a material follows the Beer-Lambert law, which states that the absorbance is proportional to the concentration of the material, the path length of the sample, and the molar absorptivity of the material at the specific wavelength of interest. Mathematically, this can be expressed as:

$$A = \varepsilon c l$$

where A is the absorbance, ε is the molar absorptivity, c is the concentration of the material, and l is the path length of the sample.

The Beer-Lambert law is widely used in analytical chemistry and biochemistry to quantify the concentration of various substances, including proteins, nucleic acids, and small molecules as shown in Figure 3.2. It is commonly used in spectroscopy techniques, such as UV-Vis spectrophotometry and fluorescence spectroscopy, to measure the absorbance of a sample at a specific wavelength.

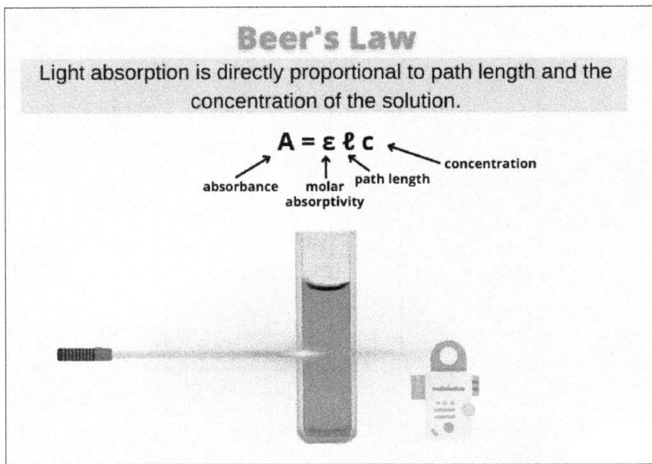

Figure 3.2 The Beer-Lambert law states that the absorption of light by a sample is directly proportional to its path length through the sample and the solution concentration.

3.3.1 Applications of Absorbance

Absorbance measurements are useful in a wide range of applications, including medical diagnostics, environmental monitoring, and industrial quality control. For example, in medical diagnostics, absorbance measurements are used to detect the presence of various biomolecules, such as glucose, cholesterol, and haemoglobin, in blood samples. In environmental monitoring, absorbance measurements are used to monitor the concentration of pollutants in air and water samples.

3.4 Fluorescence

Fluorescence is a phenomenon in which a molecule absorbs light of a specific wavelength (excitation wavelength) and subsequently emits light of a longer wavelength (emission wavelength). The emitted light is typically of lower energy and longer wavelength than the absorbed light. Fluorescence is a commonly used technique in various fields, including biochemistry, biophysics, molecular biology, and medical research. It is widely used to study the structure and function of biological molecules, such as proteins, nucleic acids, and lipids, and to detect and quantify various biomolecules and cellular components. The fluorescence process occurs when a molecule absorbs light and enters an excited state. The excited state is typically a higher-energy electronic state that is unstable and quickly relaxes to a lower-energy electronic state. This relaxation process can occur in various ways, including emission of a photon of light (fluorescence) or non-radiative relaxation, in which energy is dissipated as heat.

3.4.1 Principle of Fluorescence

The basic principle of fluorescence can be illustrated using a Jablonski diagram (Figure 3.4 (a)). A fluorescent molecule can be excited to a higher vibrational energy state (S_1^{vib} in Figure 3.4 (a)) with a light source. This excitation is most efficient at the optimal excitation wavelength of the fluorophore, which most often also corresponds to the absorption maximum. After this process of photon absorption, which happens almost instantly, the molecule quickly relaxes without radiation to the lowest vibrational energy level of S_1. This state has a much longer lifetime (~ ns) as compared to a vibrational state (~ ps), which leads to the following transition to originate only from this state. During the next transition, the molecule falls back into the ground state (S_0) either by non-radiative decay or by spontaneous emission. In the first case energy is released in the form of heat, in the second by releasing energy in the form of a photon. Since energy has been lost during the process due to the vibrational relaxation step, the emitted fluorescent photon has a lower energy than the excitation photon which leads to a red shift of the emission profile relative to the excitation profile. Quantum mechanically the exact transition leading to fluorescence are not clearly defined, but is described by a transition probability. Together with various inhomogeneous and homogeneous broadening effects, the emission profile will have a broad continuous appearance and typically resembles a shifted mirrored version of the absorption profile (Figure 3.4 (b)).

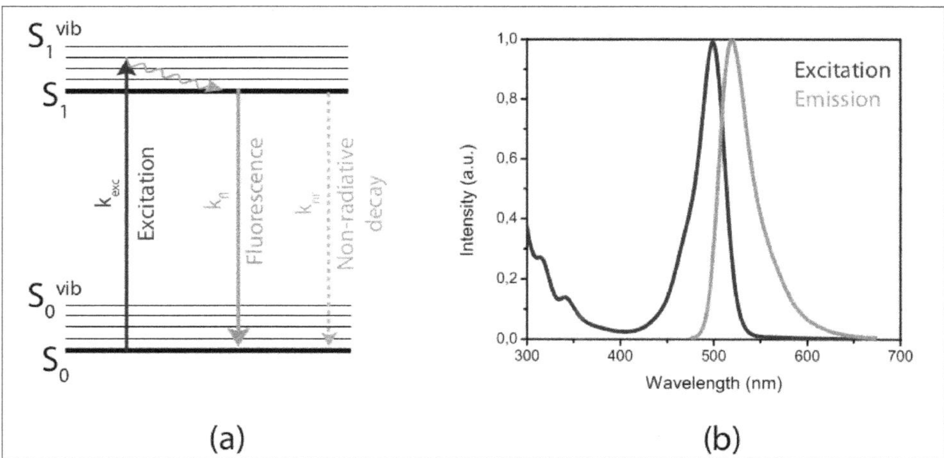

Figure 3.3 (a) Jablonski diagram illustrating the energy state transitions leading to {fluorescence} and non-radiative decay to the ground state. (b) Excitation and the red shifted emission profile of Alexa 488. Here it is clearly visible that the emission profile is red shifted and is similar to a mirrored version of the excitation profile.

The fluorescence emission spectrum of a molecule is specific to its chemical structure and is used to identify and quantify the molecule in a sample. Fluorescence measurements can be performed using a variety of techniques, including fluorescence microscopy, fluorescence spectroscopy, and fluorescence imaging.

3.4.2 Applications of Fluorescence

Fluorescence is also used in various applications, such as fluorescence-based biosensors, which use the fluorescence signal to detect and quantify the presence of specific biomolecules in a sample. Other applications include fluorescent labeling of biomolecules for imaging and tracking, and fluorescence-guided surgery for the detection and removal of cancerous tissue.

3.5 Raman Spectrum

Raman spectrum is a vibrational spectroscopy technique that provides information about the molecular structure and composition of a sample. It is based on the inelastic scattering of light by molecules, where the scattered light has a different energy and wavelength than the incident light. When a sample is illuminated with a laser, a small fraction of the scattered light undergoes a shift in frequency due to interaction with the sample molecules. This shift in frequency, known as the Raman shift, is related to the vibrational frequencies of the chemical bonds in the molecule, and is unique for each molecule as shown in Figure 3.4.

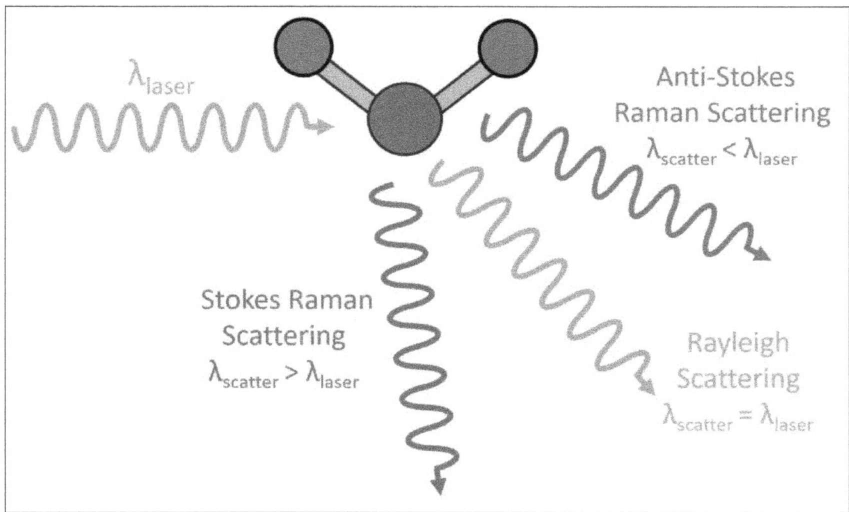

Figure 3.4 Three types of scattering processes that can occur when light interacts with a molecule

The Raman spectrum is a plot of the intensity of the scattered light as a function of the Raman shift. It contains a series of peaks that correspond to the different vibrational modes of the molecule. The Raman spectrum can be used to identify and quantify the molecules in a sample, and to study their chemical and physical properties.

3.5.1 Applications of Raman spectroscopy

Raman spectroscopy is a non-destructive and non-invasive technique that can be used to analyze a wide range of samples, including liquids, solids, and gases. It is widely used in various fields, including chemistry, materials science, biology, and forensic science.

3.5.2 Advantages of Raman spectroscopy

Raman spectroscopy has several advantages over other spectroscopy techniques. For example, it can be used to analyze samples in their natural state, without the need for sample preparation or labeling. It also has high spatial resolution and can be used to study small areas of a sample, such as individual cells or subcellular structures. Overall, Raman spectroscopy is a powerful analytical tool that provides valuable information about the structure and composition of a wide range of samples.

3.6 Colorimetric Biosensors

Colorimetric biosensors are analytical devices that use changes in colour to detect and quantify the presence of target analytes. They are based on the interaction between the analyte and a chemical or biological element, which results in a measurable colour change.[7] Colorimetric biosensors are widely used in various fields, including medical diagnostics, environmental monitoring, and food safety. They offer several advantages over other analytical techniques, such as simplicity, rapid response time, and low cost.

3.6.1 Working principle of Colorimetric biosensors

The working principle of colorimetric biosensors is based on the specific interaction between the analyte and the sensor element, which produces a color change that can be detected and measured using a simple optical system. The color change is usually caused by a chemical reaction, such as the oxidation or reduction of a chromogenic compound, or by a biological reaction, such as an enzyme-catalyzed reaction.

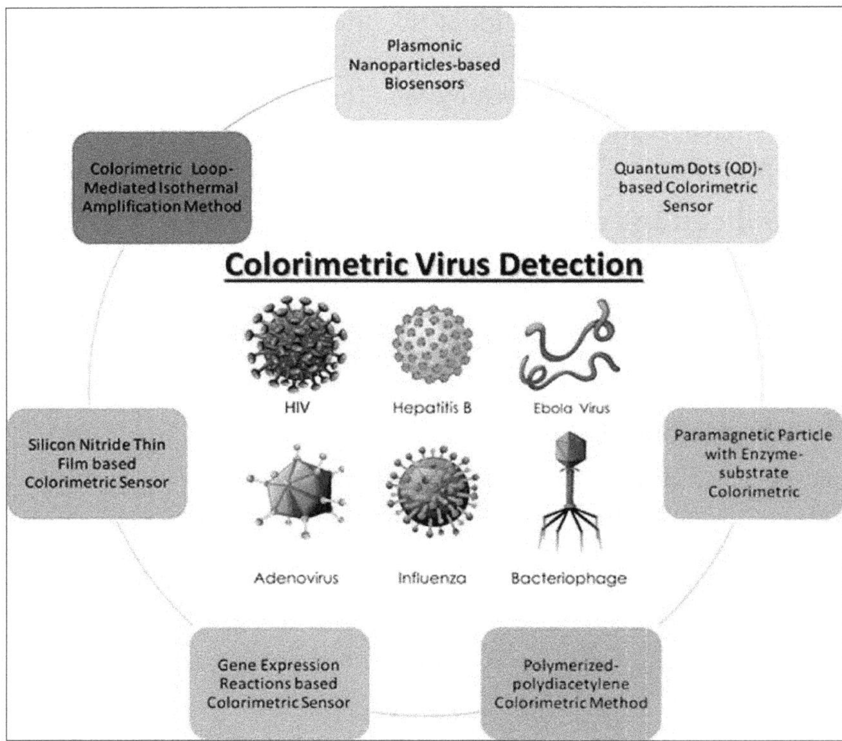

Figure 3.5 Schematic illustration for Overview of existing technologies for colorimetric virus detection.

3.6.2 Applications of Colorimetric Biosensors

Colorimetric biosensors can be designed to detect a wide range of analytes, including ions, small molecules, proteins, and nucleic acids. They can be used to detect and quantify analytes in various samples, such as blood, urine, saliva, and water.

3.6.3 Advantages of Colorimetric Biosensors

Colorimetric biosensors are often designed to be portable and user-friendly, allowing for rapid on-site analysis. They can also be integrated with mobile devices, such as smartphones, to enable remote monitoring and data analysis. Overall, colorimetric biosensors offer a simple, low-cost, and rapid analytical tool for the detection and quantification of various analytes. They have broad applications in medical diagnostics, environmental monitoring, and food safety, among other fields.

3.7 Surface Plasmon Resonance (SPR)

Surface plasmon resonance (SPR) is an optical technique used to study the interaction between biomolecules, such as proteins and nucleic acids, and their ligands or analytes. It is based on the phenomenon of surface plasmons, which are oscillations of free electrons on the surface of a thin metal film.

3.7.1 Working Principle of Surface Plasmon Resonance

In an SPR experiment, a thin metal film, typically gold or silver, is coated with a layer of biomolecules, such as antibodies or DNA. The metal film is then illuminated with a light source, such as a laser, at a specific angle as shown in Figure 3.6.

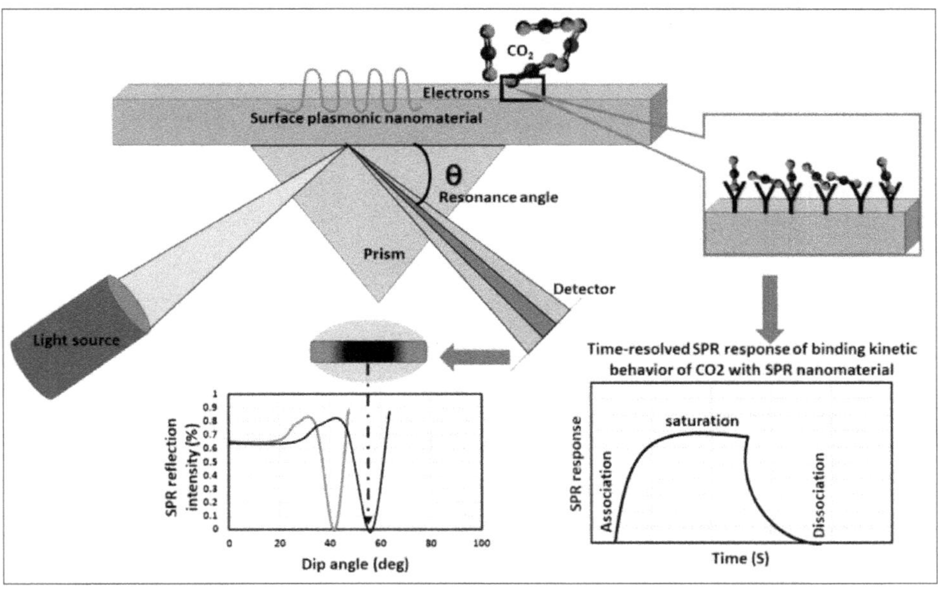

Figure 3.6 Mechanism of surface plasmon resonance (SPR) nanomaterial sensing for CO_2.

3.7.2 Applications of Surface Plasmon Resonance

SPR is a label-free and real-time technique that allows for the detection and quantification of biomolecular interactions without the need for chemical labels or fluorescent tags. It is widely used in various fields, including drug discovery, diagnostics, and biotechnology. SPR is also used for high-throughput screening of small molecule libraries for drug discovery. It allows for the rapid screening of large numbers of compounds for their binding affinity to a target protein. Overall, SPR is a powerful technique for the study of biomolecular interactions, with applications in drug discovery, diagnostics, and biotechnology.

3.8 Magnetic Biosensors

Magnetic biosensors are analytical devices that use magnetic fields to detect and quantify the presence of target analytes. They are based on the interaction between magnetic particles and the analyte, which results in a measurable change in the magnetic field.

3.8.1 Working Principle of Magnetic biosensors

The working principle of magnetic biosensors is based on the specific interaction between the analyte and the magnetic particles.

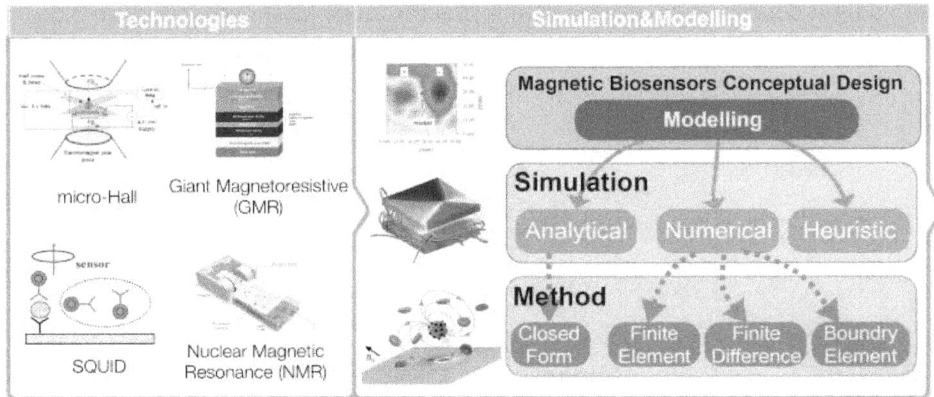

Figure 3.7 Schematic diagram representation that the magnetic biosensors: Modelling and simulation

3.8.2 Advantages of Magnetic biosensors

Magnetic biosensors offer several advantages over other analytical techniques, such as sensitivity, specificity, and rapid response time. They can also be used for the detection of a wide range of analytes, including nucleic acids, proteins, and cells. They can be designed to be portable and user-friendly, allowing for rapid on-site analysis. They can also be integrated with microfluidic systems to enable automated sample processing and analysis. Overall, magnetic biosensors offer a sensitive, specific, and rapid analytical tool for the detection and quantification of various analytes, with broad applications in medical diagnostics, environmental monitoring, and food safety, among other fields.

3.9 Nanotechnology used in Optical biosensors

Nanotechnology plays an important role in the development of optical biosensors by enabling the creation of highly sensitive and selective sensing platforms. Here are some ways in which nanotechnology is used in optical biosensors:

- **Nanoparticles:** Nanoparticles, such as gold or silver nanoparticles, can be functionalized with specific biomolecules to act as probes in optical biosensors. The nanoparticles can enhance the sensitivity and selectivity of the biosensor by increasing the surface area and providing a strong signal due to their plasmonic properties.
- **Nanowires:** Nanowires can be used as transducing elements in optical biosensors. The nanowires can be coated with biomolecules that are selective for the analyte of interest. Binding of the analyte to the biomolecules on the nanowires can produce a change in the electrical or optical properties of the nanowires, which can be detected and used to quantify the analyte.
- **Quantum dots:** Quantum dots are nanoscale semiconducting particles that emit bright and stable fluorescence. They can be conjugated with biomolecules and used as labels in optical biosensors to provide high sensitivity and selectivity.
- **Nanoporous materials:** Nanoporous materials, such as zeolites or mesoporous silica, can be used as scaffolds for the immobilization of biomolecules in optical biosensors. The high surface area and pore size of the materials can provide a large number of binding sites for the analyte, increasing the sensitivity and selectivity of the biosensor.
- **Nanoplasmonic structures**: Nanoplasmonic structures, such as nanoholes or nanorods, can be used in optical biosensors to enhance the sensitivity of the biosensor. The interaction between the analyte and the plasmonic structures can produce a strong signal that can be detected and used to quantify the analyte.

Overall, the use of nanotechnology in optical biosensors has enabled the development of highly sensitive and selective sensing platforms with broad applications in medical diagnostics, environmental monitoring, and food safety, among other fields.

3.10 Nanotechnology used in Magnetic biosensors

Nanotechnology has also played an important role in the development of magnetic biosensors by providing unique magnetic nanoparticles and functionalization techniques to improve the detection sensitivity and specificity. Here are some ways in which nanotechnology is used in magnetic biosensors:

- **Magnetic nanoparticles**: Magnetic nanoparticles, such as iron oxide nanoparticles, can be functionalized with specific biomolecules to act as probes in magnetic biosensors. The nanoparticles can provide a high surface area and improve the binding capacity of the biosensor, while also producing a measurable change in the magnetic field when they interact with the target analyte.

- **Nanoparticle clusters**: Nanoparticle clusters, formed by aggregation of magnetic nanoparticles, can be used to increase the magnetic moment and provide a stronger signal in magnetic biosensors. Additionally, the clustering of nanoparticles can enhance the specificity of the biosensor by increasing the number of binding sites and reducing the non-specific binding.
- **Nanowires:** Nanowires can be used in magnetic biosensors to improve the sensitivity and specificity of the biosensor. The nanowires can be coated with specific biomolecules, such as antibodies or DNA probes, to bind to the target analyte. The interaction between the target analyte and the nanowires produces a measurable change in the magnetic field, which can be detected and quantified.

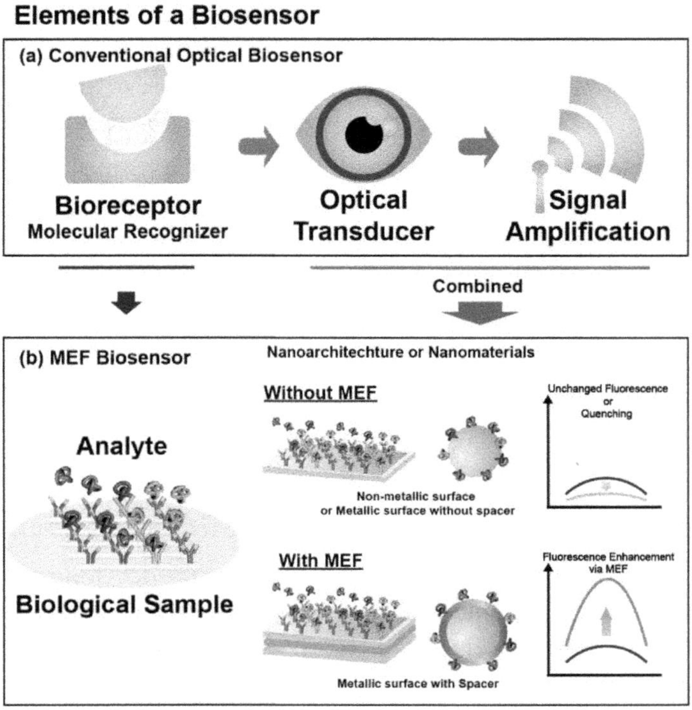

Figure 3.8 Conventional optical biosensor correlation to MEF platforms for optical biosensors [13]

- **Magnetic nanocomposites:** Magnetic nanocomposites, consisting of magnetic nanoparticles embedded in a polymeric or metallic matrix, can be used in magnetic biosensors to improve the stability and specificity of the biosensor. The matrix can protect the magnetic nanoparticles from degradation and improve their binding capacity, while also providing a suitable platform for immobilizing the biomolecules.
- **Magnetic microbeads**: Magnetic microbeads, consisting of magnetic nanoparticles embedded in a polymeric or silica matrix, can be used in magnetic biosensors to improve the sensitivity and specificity of the biosensor. The microbeads can be functionalized with specific biomolecules to bind to the target analyte, and their magnetic properties can provide a measurable change in the magnetic field when they interact with the analyte.

Overall, the use of nanotechnology in magnetic biosensors has enabled the development of highly sensitive and selective sensing platforms with broad applications in medical diagnostics, environmental monitoring, and food safety, among other fields [14].

References

Dey, D., and T. Goswami. "Optical biosensors: a revolution towards quantum nanoscale electronics device fabrication." Journal of Biomedicine and Biotechnology 2011 (2011).

Mayerhöfer, Thomas G.; Pahlow, Susanne; Popp, Jürgen (2020). "The Bouguer-Beer-Lambert Law: Shining Light on the Obscure". ChemPhysChem. 21: 2031. doi:10.1002/cphc.202000464

Beer, August (1852). "Bestimmung der Absorption des rothen Lichts in farbigen Flüssigkeiten" (Determination of the absorption of red light in colored liquids)." Annalen der Physik und Chemie. 162 (5): 78–88. doi:10.1002/andp.18521620505

Britannica, The Editors of Encyclopaedia. "fluorescence". Encyclopedia Britannica, 28 Feb. 2023, https://www.britannica.com/science/fluorescence. Accessed 26 March 2023.

Raman, C., Krishnan, K. A New Type of Secondary Radiation. Nature 121, 501–502 (1928). https://doi.org/10.1038/121501c0

Daniel Němeček, George J. Thomas, Chapter 16 - Raman Spectroscopy of Viruses and Viral Proteins, Editor(s): Jaan Laane, Frontiers of Molecular Spectroscopy, Elsevier, 2009, Pages 553-595, ISBN 9780444531759, https://doi.org/10.1016/B978-0-444-53175-9.00016-7.

Victoria Xin Ting Zhao, Ten It Wong, Xin Ting Zheng, Yen Nee Tan, Xiaodong Zhou, Colorimetric biosensors for point-of-care virus detections, Materials Science for Energy Technologies, Volume 3, 2020, Pages 237-249, ISSN 2589-2991, https://doi.org/10.1016/j.mset.2019.10.002.

Wang, Qi, et al. "Research advances on surface plasmon resonance biosensors." Nanoscale 14.3 (2022): 564-591.

Rezk, Marwan & Sharma, Jyotsna & Gartia, Manas Ranjan. (2020). Nanomaterial-Based CO2 Sensors. Nanomaterials. 10. 2251. 10.3390/nano10112251.

Vahid Nabaei, Rona Chandrawati, Hadi Heidari, Magnetic biosensors: Modelling and simulation, Biosensors and Bioelectronics, Volume 103, 2018, Pages 69-86, ISSN 0956-5663, https://doi.org/10.1016/j.bios.2017.12.023.

Haun, Jered B., et al. "Magnetic nanoparticle biosensors." Wiley Interdisciplinary Reviews: Nanomedicine and Nanobiotechnology 2.3 (2010): 291-304.

Soler, Maria, and Laura M. Lechuga. "Boosting cancer immunotherapies with optical biosensor nanotechnologies." (2019).

Jeong, Yoon & Kook, Yun-min & Lee, Kangwon & Koh, Won-Gun. (2018). Metal Enhanced Fluorescence (MEF) for Biosensors: General Approaches and A Review of Recent Developments. Biosensors and Bioelectronics. 111. 10.1016/j.bios.2018.04.007.

Jianrong, Chen, et al. "Nanotechnology and biosensors." Biotechnology advances 22.7 (2004): 505-518.

Chapter-4

TRANSDUCERS & BIO-RECOGNITION ELEMENTS

Introduction

Transducers and bio-recognition elements are two important components in biosensors, which are devices that use biological molecules to detect the presence of a target analyte. A transducer is a device that converts one form of energy into another. In the context of biosensors, a transducer converts the biological signal generated by the recognition element into a measurable signal. This measurable signal can be electrical, optical, thermal, or mechanical. Examples of transducers used in biosensors include electrochemical sensors, optical sensors, and piezoelectric sensors.

Bio-recognition elements, also known as biological recognition elements, are the components of a biosensor that specifically bind to the target analyte. These elements are usually biomolecules such as enzymes, antibodies, or DNA/RNA sequences that can recognize and bind to the target analyte with high selectivity and sensitivity. The recognition element is immobilized onto the surface of the transducer, where it can interact with the target analyte. In summary, transducers and bio-recognition elements work together to convert the presence of a target analyte into a measurable signal that can be detected and quantified.

4.2 Transducers

Transducers are devices that convert one form of energy into another. They are used in a wide range of applications, from everyday devices such as microphones and speakers to more specialized applications such as sensors and measurement systems as shown in Figure 4.1.

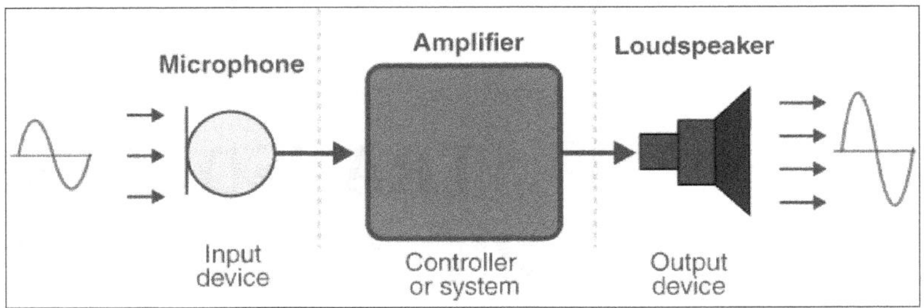

Figure 4.1 Schematic illustration representation that the working mechanism of transducer

In the context of sensors, a transducer is a device that converts a physical, chemical or biological signal into an electrical signal that can be measured and analysed. Some common examples of transducers used in sensors include:

- **Strain gauges** - These transducers measure the strain or deformation in a material by converting it into a change in electrical resistance.
- **Thermocouples** - These transducers measure temperature by converting the temperature difference between two metals into an electrical voltage.
- **Pressure sensors** - These transducers measure pressure by converting it into an electrical signal.
- **Accelerometers** - These transducers measure acceleration by converting it into an electrical signal.
- **Piezoelectric sensors** - These transducers convert mechanical stress or pressure into an electrical signal.

There are many other types of transducers used in various applications, but the underlying principle is the same: they convert one form of energy into another that can be more easily measured and analyzed.

4.3 Bio-recognition Elements

Bio-recognition elements, also known as biological recognition elements, are molecules that specifically recognize and bind to a particular target molecule or analyte. These molecules are typically biological in nature, such as enzymes, antibodies, or nucleic acids. In biosensors, bio-recognition elements are used as a way to selectively detect and quantify specific analytes in complex biological samples, such as blood or urine. For example, an antibody-based biosensor may use an antibody as the bio-recognition element that specifically binds to a target antigen, while other non-specific molecules are excluded from binding as shown in Figure 4.2.

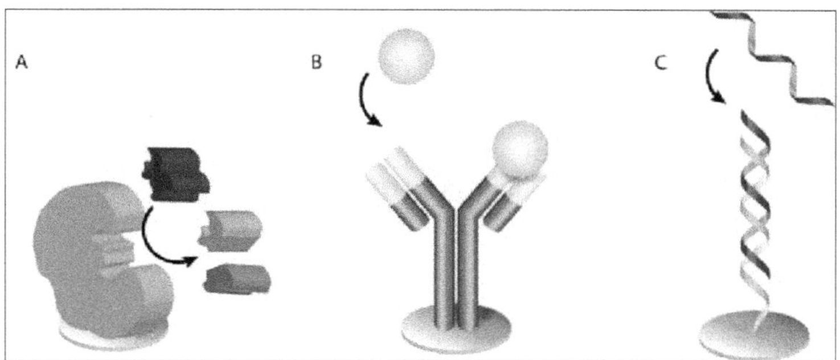

Figure 4.2 Main types of biosensors depending on the biorecognition element: (a) enzymatic biosensor (catalytic), (b) immunosensor (affinity) and (c) DNA biosensor (affinity).

Enzymes are another common type of bio-recognition element used in biosensors. They can specifically bind to and catalyze a chemical reaction with their substrate, producing a detectable signal that can be measured by the transducer. For example, a glucose biosensor may use the enzyme glucose oxidase as the bio-recognition element to specifically recognize and oxidize glucose, generating a measurable electrical signal. In addition to antibodies and enzymes, other types of bio-recognition elements include aptamers, which are short DNA or RNA sequences that can specifically bind to a target molecule, and receptors, which are proteins that bind to specific ligands or molecules. The choice of bio-recognition element depends on the specific application and the characteristics of the target analyte, as different molecules may have different affinities, sensitivities, and specificities for different targets.

4.4 Enzymes

Enzymes are biological molecules, typically proteins, that catalyse or speed up chemical reactions in living organisms. They play a vital role in many biological processes, including digestion, metabolism, and cellular signalling as shown in Figure 4.3. Enzymes are highly specific, recognizing and binding to specific substrates or molecules to catalyse a specific reaction.

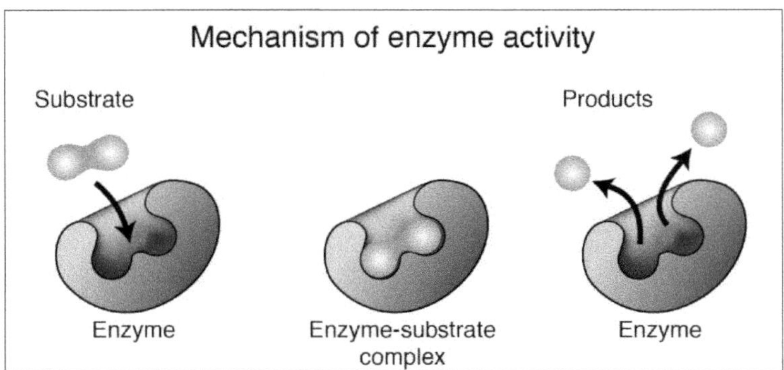

Figure 4.3 Schematic diagram representation that the mechanism of enzyme activity

4.4.1 Properties of Enzymes

Enzymes have several important properties that make them useful in various applications, including in biosensors. These properties include:

- **High specificity:** Enzymes can recognize and bind to specific substrates with high specificity, allowing for the selective detection of target molecules.
- **High sensitivity:** Enzymes can catalyse reactions with high efficiency, allowing for the detection of low concentrations of target molecules.
- **Stability:** Enzymes are stable under a wide range of conditions, including high and low temperatures and pH values.
- **Reproducibility:** Enzymatic reactions are highly reproducible, allowing for consistent and accurate measurements.

Enzymes are commonly used as bio-recognition elements in biosensors, where they can be used to selectively detect and quantify specific analytes. For example, glucose oxidase is an enzyme commonly used in glucose biosensors to catalyse the oxidation of glucose, producing a measurable signal that can be detected by a transducer. Other enzymes used in biosensors include cholinesterase, lactate dehydrogenase, and acetylcholinesterase, among others.[4] Enzymes are also used in many other applications, such as in the food industry for the production of cheese and bread, and in medicine for the treatment of diseases.

4.5 Antibodies

Antibodies, also known as immunoglobulins, are specialized proteins produced by the immune system to identify and neutralize harmful pathogens such as bacteria, viruses, and other foreign substances. Antibodies are Y-shaped molecules that consist of four protein chains: two heavy chains and two light chains. Each antibody has a unique region at the tip of the Y-shaped molecule called the antigen-binding site, which recognizes and binds to specific antigens on the surface of pathogens.

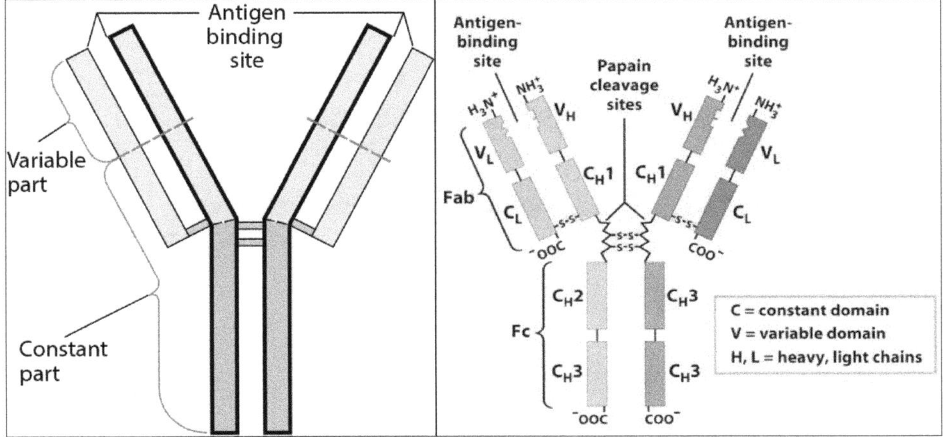

Figure 4.4 Schematic illustration representation that the Y Structure of Antibodies

When a pathogen enters the body, the immune system produces antibodies that are specific to that pathogen's antigens. The antibodies bind to the antigens and mark the pathogen for destruction by other immune cells, such as phagocytes and natural killer cells. Some antibodies can also neutralize pathogens by blocking their ability to enter or infect cells. Antibodies can also provide immunity to future infections by "remembering" how to produce the specific antibody needed to fight off a previously encountered pathogen [5]. There are several different types of antibodies, each with a specific role in the immune system. For example, IgM antibodies are the first antibodies produced in response to an infection, while IgG antibodies provide long-term protection against specific pathogens.

4.6 Receptors

Receptors are specialized proteins that are found on the surface of cells or inside cells. They play a key role in the communication between cells and in the regulation of cellular processes. Receptors can detect and respond to various types of signals, such as hormones, neurotransmitters, and other signaling molecules. When a signaling molecule binds to a receptor on the surface of a cell, it can trigger a series of events inside the cell that lead to a specific response. For example, the binding of a hormone to its receptor can activate a signaling pathway that leads to the production of a second messenger molecule, which in turn can activate various enzymes and other proteins that regulate cellular processes as shown in Figure 4.5.

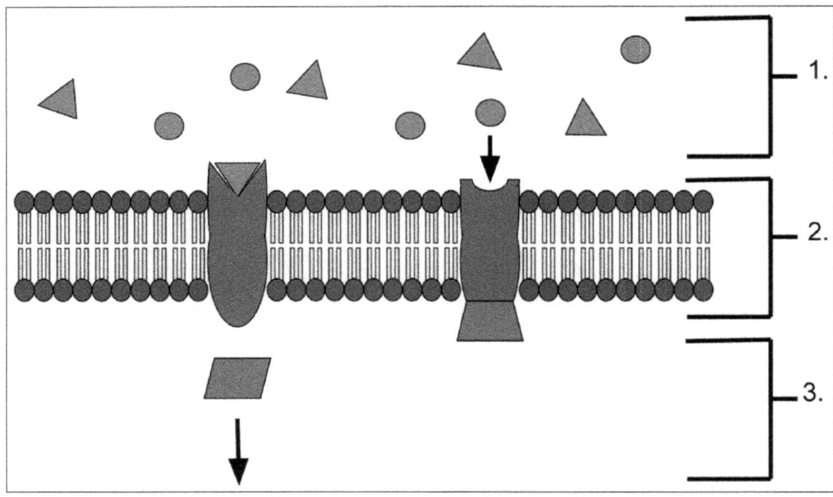

Figure 4.5 Membrane receptors (1) Ligands, located outside the cell (2) Ligands connect to specific receptor proteins based on the shape of the active site of the protein (3) The receptor releases a messenger once the ligand has connected to the receptor.

There are several different types of receptors, including ion channel receptors, G protein-coupled receptors, and enzyme-linked receptors. Ion channel receptors are involved in the transmission of electrical signals between cells, while G protein-coupled receptors are involved in a wide range of physiological processes, including neurotransmission, hormone signaling, and the regulation of cardiovascular function. Enzyme-linked receptors play a role in the regulation of cell growth, differentiation, and survival.[6] Receptors are important targets for drugs and other therapeutic agents. Many drugs work by binding to specific receptors and modulating their activity, either by enhancing or inhibiting the signaling pathway that they activate. Understanding the structure and function of receptors is therefore essential for the development of new drugs and therapies for various diseases and conditions.

4.7 Nucleic Acids

Nucleic acids are biological macromolecules that store, transmit, and express genetic information in living organisms. There are two main types of nucleic acids: deoxyribonucleic acid (DNA) and ribonucleic acid (RNA) as representation that the Figure 4.6.

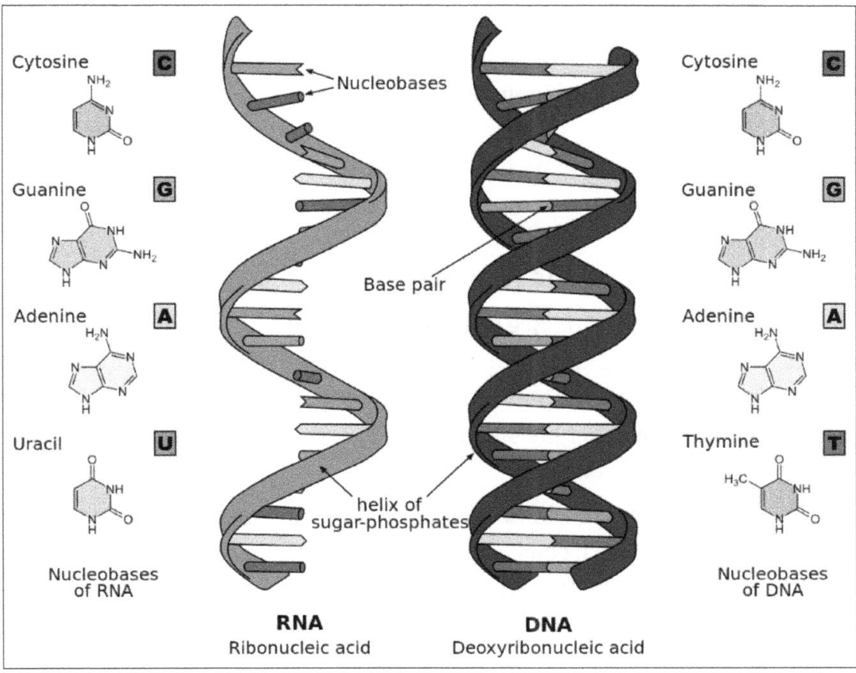

Figure 4.6 Schematic illustration of Nucleic acids RNA and DNA

4.7.1 Deoxyribonucleic Acid (DNA)

DNA is the genetic material that encodes the instructions for the development, function, and reproduction of all living organisms. It is a long, double-stranded molecule that is made up of four types of nucleotides: adenine (A), cytosine (C), guanine (G), and thymine (T). The sequence of these nucleotides, known as the genetic code, determines the sequence of amino acids in proteins, which are the building blocks of cells and tissues.

4.7.2 Ribonucleic Acid (RNA)

RNA is a single-stranded molecule that is involved in the translation of genetic information into proteins. It is made up of the same four types of nucleotides as DNA, except that thymine is replaced by uracil (U). There are three main types of RNA: messenger RNA (mRNA), transfer RNA (tRNA), and ribosomal RNA (rRNA). mRNA carries the genetic code from DNA to the ribosomes, where it is translated into proteins. tRNA brings amino acids to the ribosomes, where they are assembled into proteins. rRNA is a structural component of the ribosome, which is the molecular machine that carries out protein synthesis. Nucleic acids are essential for life and play a critical role in a wide range of biological processes, including cell division, growth, and differentiation, as well as the response to environmental stresses and the regulation of gene expression. Understanding the structure and function of nucleic acids is important for advancing our knowledge of genetics and developing new treatments for genetic disorders and diseases.

4.8 Biosensor

4.8.1 Introduction of Biosensors

A biosensor is an analytical device that combines a biological sensing element with a transducer to detect and convert a biological response into an electrical or optical signal. Biosensors are commonly used to detect the presence or concentration of biological molecules such as enzymes, antibodies, or DNA, as well as various environmental contaminants.

The biological sensing element of a biosensor can be made up of various materials such as enzymes, cells, or nucleic acids, and can be used to specifically detect a target analyte. The transducer then converts the biological response into a measurable signal, which can be displayed on a monitor or recorded for further analysis.

4.8.2 Fabrication of a Biosensor

The fabrication of a biosensor typically involves several steps, including the selection and preparation of the biological sensing element, the design and manufacture of the transducer, and the integration of the sensing element and transducer into a functional device. The following is a general overview of the biosensor fabrication process:

- **Selection of the biological sensing element:** The biological sensing element is typically chosen based on its ability to specifically recognize and bind to the target analyte. Common sensing elements include enzymes, antibodies, nucleic acids, and whole cells. The sensing element may need to be purified, immobilized, or otherwise prepared for use in the biosensor.
- **Design and manufacture of the transducer:** The transducer is the component of the biosensor that converts the biological response into a measurable signal. Depending on the application, the transducer may be an electrochemical, optical, or piezoelectric device. The transducer is designed to be compatible with the sensing element and to provide a sensitive and specific response to the target analyte.

- **Integration of the sensing element and transducer:** The sensing element and transducer are integrated into a functional device. This may involve the immobilization of the sensing element onto the transducer surface, or the use of a membrane or other interface to facilitate the interaction between the sensing element and the analyte.
- **Testing and optimization:** The biosensor is tested and optimized to ensure that it provides accurate and reliable results. This may involve calibration of the device, testing with known analyte concentrations, and optimization of the operating conditions.

4.8.3 Applications of Biosensors

Biosensors have a wide range of applications, including in healthcare for medical diagnostics, in food safety for detecting contaminants, in environmental monitoring for detecting pollutants, and in industrial processes for monitoring production processes. The development of biosensors has led to the development of new diagnostic tools that are faster, more accurate, and more sensitive than traditional methods, with the potential for widespread use in a variety of fields.

4.9 Immobilization of Biomolecules

Immobilization of biomolecules refers to the process of attaching or fixing them to a surface or matrix, typically in a stable and functional manner, for use in various applications including biosensors, drug delivery, and biocatalysis. The immobilization process can be achieved through various methods, including physical adsorption, covalent bonding, and entrapment.

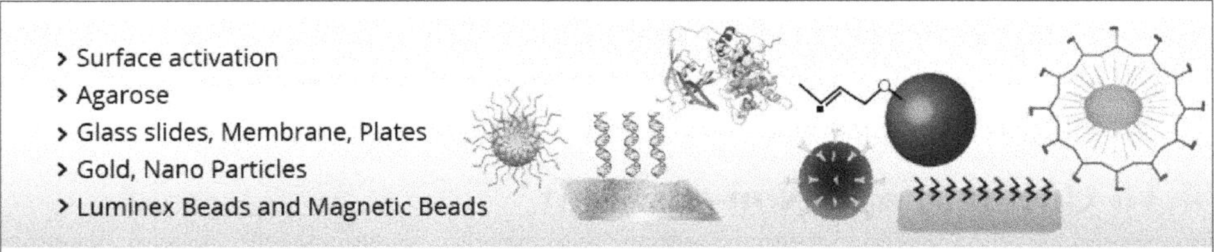

Figure 4.7 schematic illustration that the immobilization of biomolecules

4.9.1 Methods of Immobilization

4.9.1.1 Physical Adsorption

In this method, the biomolecules are attached to the surface or matrix through weak interactions such as van der Waals forces, hydrophobic interactions, or hydrogen bonding. Physical adsorption is a simple and fast method but it may result in poor stability and low reproducibility.

4.9.1.2 Covalent Bonding

This method involves the formation of a strong chemical bond between the biomolecule and the surface or matrix, typically through the use of a coupling agent or linker. Covalent bonding offers better stability and reproducibility compared to physical adsorption, but it can be more complex and time-consuming.

4.9.1.3 Crosslinking

Crosslinking involves the formation of covalent bonds between the biomolecule and a crosslinking agent to create a stable network. Crosslinking can be achieved through various methods, including glutaraldehyde crosslinking and carbodiimide crosslinking.

4.9.1.4 Entrapment

In this method, the biomolecules are encapsulated within a porous matrix or membrane, typically through the use of polymers. Entrapment provides good stability and high loading capacity, but it may result in limited access to the biomolecule and slow diffusion of the analyte.

4.9.1.5 Affinity Immobilization

This method involves the use of specific interactions between the biomolecule and the surface or matrix, such as biotin-streptavidin or antibody-antigen interactions. Affinity immobilization can provide high specificity and sensitivity, but it requires the availability of specific ligands or antibodies. The choice of immobilization method depends on various factors, including the type and size of the biomolecule, the properties of the surface or matrix, and the intended application. Immobilization of biomolecules is an important step in the fabrication of biosensors, as it allows for the specific detection of target analytes and enhances the stability and sensitivity of the sensor.

4.10 Covalent and Non-covalent

Covalent and non-covalent bonds are two types of chemical bonds that differ in their strength and the nature of the interaction between the atoms or molecules involved.

4.10.1 Covalent Bonds

Covalent bonds involve the sharing of electrons between atoms to form a stable molecular bond. In a covalent bond, the atoms share electrons to achieve a stable electron configuration, which is typically achieved by filling their outermost valence shells. Covalent bonds are strong and require a significant amount of energy to break. They are typically formed between atoms of non-metallic elements, such as carbon, hydrogen, oxygen, and nitrogen.

4.10.2 Non-covalent Bonds

Non-covalent bonds are weaker than covalent bonds and involve interactions between charged or polar molecules. There are four types of non-covalent bonds: hydrogen bonds, electrostatic interactions, van der Waals forces, and hydrophobic interactions. Non-covalent bonds play important roles in biological systems, where they are responsible for many molecular interactions such as protein-protein interactions, enzyme-substrate interactions, and DNA base pairing. In biomolecules and biotechnology, both covalent and non-covalent bonding are used for various purposes. Covalent bonding is often used for the immobilization of biomolecules onto surfaces or matrices for various applications, such as biosensors, drug delivery systems, and biocatalysts. Non-covalent interactions are also important in biomolecule interactions, such as protein-protein interactions and enzyme-substrate interactions, and in the folding and stabilization of biomolecules such as proteins and nucleic acids.

4.11 Self-assembly Method

Self-assembly is a process in which molecules or components spontaneously organize themselves into ordered structures or patterns, without external intervention. Self-assembly is a powerful technique in materials science and nanotechnology, as it allows the creation of complex structures and patterns with high precision and control. There are several methods of self-assembly, including:

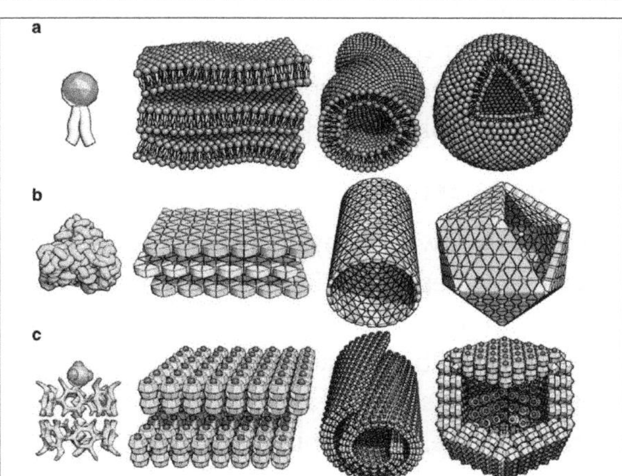

Figure 4.9 Schematic illustration that the self-assembly of (a) lipids (b) proteins (c) SDS-cyclodextrin complexes. SDS is a surfactant with a hydrocarbon tail (yellow) and a SO_4 head (blue and red), while cyclodextrin is a saccharide ring (green C and red O atoms).

- **Supramolecular self-assembly:** Supramolecular self-assembly involves the formation of non-covalent interactions between molecules, such as hydrogen bonds, electrostatic interactions, van der Waals forces, and hydrophobic interactions, to form ordered structures or patterns. This method is commonly used in the design of supramolecular materials, such as molecular gels, liquid crystals, and porous materials.
- **Langmuir-Blodgett method**: The Langmuir-Blodgett method involves the transfer of a monolayer of molecules or nanoparticles from the surface of a liquid onto a solid substrate. This method is commonly used to create thin films with controlled thickness and orientation, which can be used in electronic, optical, and sensor applications.
- **Dip-pen nanolithography**: Dip-pen nanolithography is a technique in which a sharp tip is used to deposit molecules or nanoparticles onto a substrate to create patterns with high resolution and control. This method is commonly used in the fabrication of nanoelectronics, biosensors, and microfluidic devices.
- **Template-directed self-assembly:** Template-directed self-assembly involves the use of a pre-existing template to guide the self-assembly of molecules or nanoparticles into a desired pattern or structure. This method is commonly used in the fabrication of nanowires, nanotubes, and other complex structures.

Self-assembly is a versatile technique with many potential applications in materials science, nanotechnology, and biotechnology. By controlling the properties of the molecules or components involved, it is possible to create a wide range of complex structures and patterns with high precision and control, which can be used in a variety of applications [11].

4.12 Physical Adsorption

Physical adsorption, also known as physisorption or van der Waals adsorption, is a process in which molecules or particles are adsorbed onto a surface through weak, non-covalent interactions, such as van der Waals forces, hydrogen bonding, or electrostatic interactions. Physical adsorption is a reversible process and the adsorbed molecules or particles can be easily desorbed by changing the conditions, such as temperature, pressure, or the composition of the surrounding medium.

4.12.1 Applications of Physical Adsorption

Physical adsorption is commonly used in various fields, including materials science, chemistry, and biology. Some common applications of physical adsorption are:

- **Gas separation and purification**: Physical adsorption is used in gas separation and purification processes, where the target gas is adsorbed onto a surface with high adsorption capacity, such as activated carbon or zeolites.
- **Catalysis:** Physical adsorption is often used in heterogeneous catalysis, where the catalyst is adsorbed onto a surface and the reactants are adsorbed onto the catalyst surface, promoting the reaction rate.
- **Chromatography:** Physical adsorption is used in chromatography techniques, such as gas chromatography and liquid chromatography, where the sample is adsorbed onto a stationary phase and the separation is achieved based on the differential adsorption of different components.
- **Surface modification:** Physical adsorption can be used to modify the surface properties of materials, such as wettability, surface charge, or biocompatibility, by adsorbing molecules with specific functional groups onto the surface.

Overall, physical adsorption is a versatile technique that can be used for various applications, especially in separation, purification, and modification of materials and surfaces.

References

Bojańczyk, Mikołaj. "Transducers with origin information." Automata, Languages, and Programming: 41st International Colloquium, ICALP 2014, Copenhagen, Denmark, July 8-11, 2014, Proceedings, Part II 41. Springer Berlin Heidelberg, 2014.

Sánchez, César. (2016). Nanophotonic Biosensors for Deciphering Gene Regulation Pathways. 10.13140/RG.2.1.2034.3447.

Morales, Marissa A., and Jeffrey Mark Halpern. "Guide to selecting a biorecognition element for biosensors." Bioconjugate chemistry 29.10 (2018): 3231-3239.

Atalah, Joaquín, et al. "Thermophiles and the applications of their enzymes as new biocatalysts." Bioresource technology 280 (2019): 478-488.

Forthal, Donald N. "Functions of antibodies." Microbiology spectrum 2.4 (2014): 2-4.

Cuatrecasas, "Membrane receptors." Annual review of biochemistry 43.1 (1974): 169-214.

Chargaff, Erwin, ed. The nucleic acids. Elsevier, 2012.

Saxena das. "Nanomaterials towards fabrication of cholesterol biosensors: Key roles andsign approaches." Biosensors and Bioelectronics 75 (2016): 196-205.

Kandimalla, Vivek Babu, Vijay Shyam Tripathi, and Huangxian Ju. "Immobilization of biomolecules in sol–gels: biological and analytical applications." Critical Reviews in Analytical Chemistry 36.2 (2006): 73-106.

Boott, Charlotte E., Ali Nazemi, and Ian Manners. "Synthetic covalent and non-covalent 2D materials." Angewandte Chemie International Edition 54.47 (2015): 13876-13894.

Xu, Zongwei, et al. "A review on colloidal self-assembly and their applications." Current Nanoscience 12.6 (2016): 725-746.

Thommes, Matthias, and Katie A. Cychosz. "Physical adsorption characterization of nanoporous materials: progress and challenges." Adsorption 20.2-3 (2014): 233-250.

Chapter-5

APPLICATION OF BIOSENSORS IN AGRICULTURE

Introduction

Biosensors are analytical devices that combine a biological recognition element with a transducer to convert a biological response into an electrical or optical signal that can be measured and quantified. Biosensors have a wide range of applications in various fields, including agriculture. Here are some examples of the application of biosensors in agriculture:

- **Detection of pesticides and herbicides:** Biosensors can be used to detect the presence of pesticides and herbicides in soil, water, and crops. The biosensor can detect the specific chemical compound and provide real-time results, allowing farmers to take appropriate measures to reduce the use of chemicals or switch to safer alternatives.
- **Monitoring soil quality**: Biosensors can be used to monitor soil quality by measuring the nutrient content, pH, and other parameters that affect plant growth. This information can help farmers to optimize their fertilizer use and improve crop yields.
- **Disease detection:** Biosensors can be used to detect plant diseases caused by bacteria, fungi, and viruses. Early detection of plant diseases can help farmers to prevent the spread of the disease and take appropriate measures to treat infected plants.
- **Quality control:** Biosensors can be used to monitor the quality of agricultural products, such as fruits, vegetables, and meat. Biosensors

can detect the presence of harmful contaminants, such as bacteria or chemicals, and help ensure that the products are safe for consumption.

- **Detection of pathogens in livestock:** Biosensors can be used to detect pathogens in livestock, such as bacteria, viruses, and parasites. This can help farmers to prevent the spread of diseases and improve animal health.

Overall, biosensors have the potential to revolutionize the way agriculture is practiced by providing real-time data on soil quality, crop health, and food safety. They can help farmers to make informed decisions about their farming practices, reduce the use of harmful chemicals, and ensure the safety and quality of agricultural products.

5.2 Agriculture

Agriculture is the practice of cultivating soil, plants, and livestock to produce food, fiber, and other products to sustain human life as shown in Figure 5.1. Agriculture has been a fundamental aspect of human civilization, and it has evolved over time to become more sophisticated and efficient. Today, agriculture is a vast industry that involves a range of practices, from small-scale subsistence farming to large-scale commercial farming.

Figure 5.1 Photograph image representation that the various types of agriculture sectors

The agriculture industry is divided into several sectors, including crop production, livestock production, aquaculture, and forestry. Crop production involves growing crops such as grains, fruits, vegetables, and nuts. Livestock production involves raising animals for meat, milk, eggs, or other products. Aquaculture involves farming fish, shellfish, and other aquatic organisms. Forestry involves managing and harvesting trees for wood and other products.

Agriculture also involves a range of practices and technologies to improve efficiency and sustainability, including irrigation, fertilization, pest control, crop breeding, genetic engineering, precision agriculture, and more. The goal of modern agriculture is to produce food and other products efficiently while minimizing environmental impacts and ensuring sustainability for future generations.[2] Overall, agriculture is a critical industry that plays a crucial role in providing food and other products to sustain human life. With the global population expected to reach 9.7 billion by 2050, agriculture will continue to play a vital role in meeting the growing demand for food and other products.

5.3 Soil and Water Analysis

Biosensors can be used for soil and water analysis to detect various analytes such as nutrients, heavy metals, organic pollutants, and pathogens. The application of biosensors in soil and water analysis has several advantages, including rapid, sensitive, and selective detection, minimal sample preparation, and real-time monitoring.

5.3.1 Soil Analysis

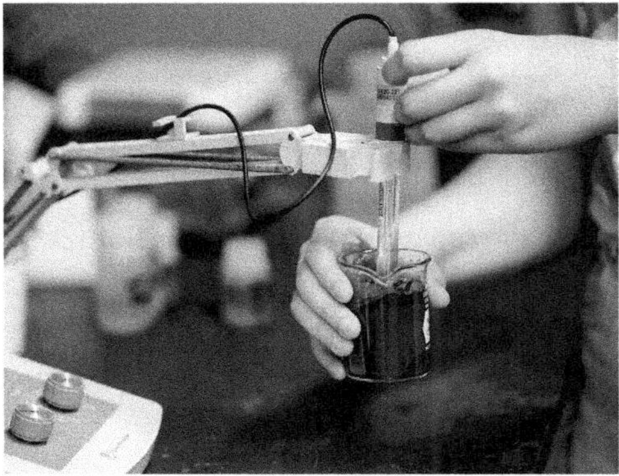

Figure 5.2 Photograph image of soil analysis test using Biosensors

In soil analysis, biosensors can be used to detect the concentration of various nutrients such as nitrogen, phosphorus, and potassium, which are essential for plant growth. Biosensors can also detect the presence of heavy metals in soil, which can be harmful to plant growth and human health as shown in Figure 5.2. Additionally, biosensors can be used to detect soil-borne pathogens, such as bacteria and fungi, which can cause diseases in plants.

5.3.2 Water Analysis

In water analysis, biosensors can be used to detect various analytes, including heavy metals, organic pollutants, and pathogens as shown in Figure 5.3. Heavy metals in water can be harmful to human health, and biosensors can detect them at very low concentrations. Organic pollutants, such as pesticides and herbicides, can also be detected using biosensors. Biosensors can also be used to detect waterborne pathogens, such as bacteria and viruses, which can cause waterborne diseases.

Figure 5.3 Photograph image of water analysis using Biosensors

Overall, biosensors have the potential to revolutionize soil and water analysis by providing rapid, sensitive, and selective detection of various analytes. This information can help farmers and environmental scientists to make informed decisions about soil and water management, ensuring the sustainability and safety of agricultural production and the environment.

5.4 Pesticide Residues

Pesticide residues are chemicals left on or in crops, soil, or water after the application of pesticides. These residues can be harmful to human health and the environment if they are not properly managed. Biosensors can be used to detect pesticide residues in various matrices, including food, water, soil, and air. Biosensors can detect the specific chemical compounds present in pesticide residues, providing rapid and sensitive detection as shown in Figure 5.4. Some biosensors can detect multiple pesticide residues simultaneously, which can save time and reduce costs. Additionally, biosensors can be used for on-site detection, which eliminates the need for samples to be sent to a laboratory for analysis, saving time and reducing the risk of sample contamination.[5]

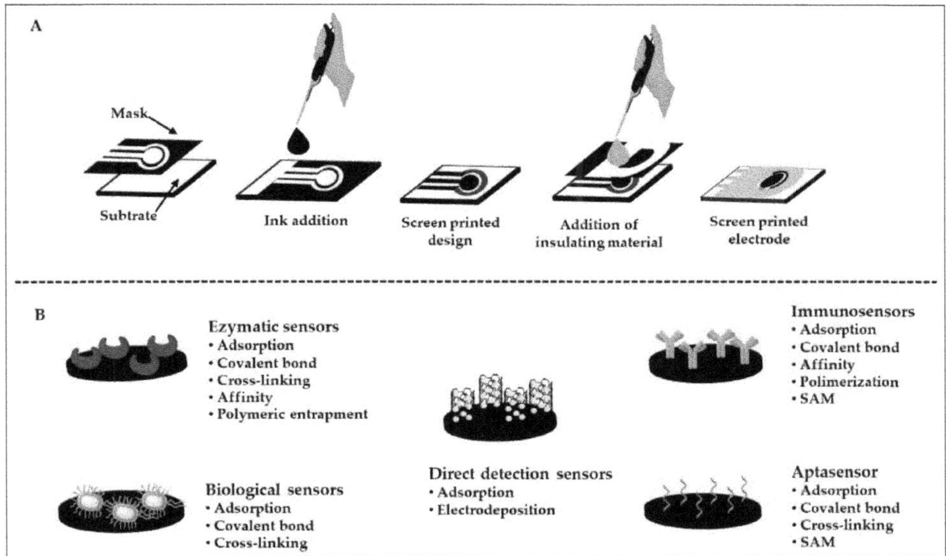

Figure 5.4 (a) Stages of the manufacturing process of screen-printed electrodes (SPEs), (b) Main methods used for the immobilization of (bio)receptors on the working electrode of SPEs for pesticides detection (SAM: Self Assembled Monolayers).

5.4.1 Use of Biosensors for Pesticide Residue Detection

The use of biosensors for pesticide residue detection has several advantages over traditional methods, such as gas chromatography and liquid chromatography as shown in Figure 5.5. Biosensors are more cost-effective, require minimal sample preparation, and can provide real-time results. Furthermore, biosensors are more environmentally friendly since they generate less waste than traditional methods.

Overall, biosensors have the potential to revolutionize the way pesticide residues are detected and managed. By providing rapid, sensitive, and selective detection, biosensors can help ensure the safety of food and the environment while reducing the use of harmful chemicals in agriculture.

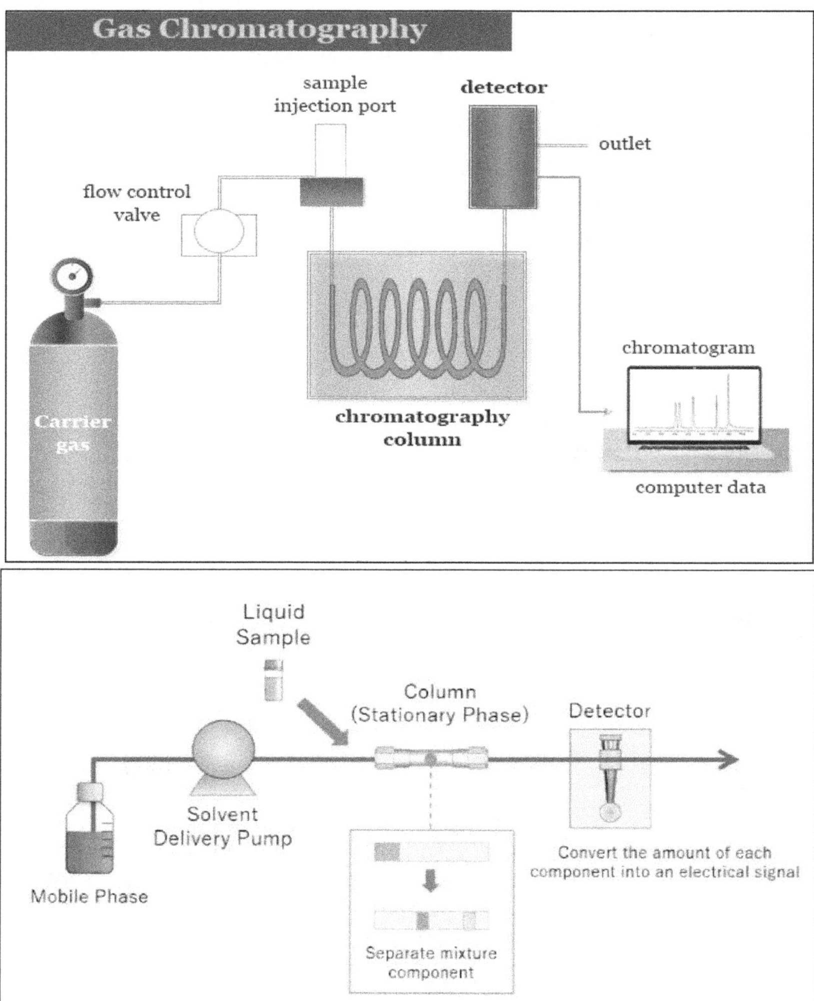

Figure 5.5 (a) Overview of Gas chromatography (GC) and (b) Overview of High-Performance Liquid Chromatography (HPLC).

5.5 Early Detection of Plant Diseases

Early detection of plant diseases is critical for preventing the spread of diseases and minimizing crop losses. Biosensors can be used for early detection of plant diseases by detecting the specific biomolecules produced by pathogens or the host plant in response to pathogen attack. Biosensors can detect these biomolecules at very low concentrations, allowing for the early detection of plant diseases before visible symptoms appear.

5.5.1 Biosensors in Early Detection of Plant Diseases

Biosensors can be designed to detect various biomolecules associated with plant diseases, such as proteins, enzymes, nucleic acids, and secondary metabolites. For example, biosensors can be designed to detect pathogen-specific proteins or DNA sequences, allowing for the specific identification of the pathogen causing the disease. Biosensors can also detect plant-derived biomolecules, such as phytohormones and volatile organic compounds, which are produced in response to pathogen attack.

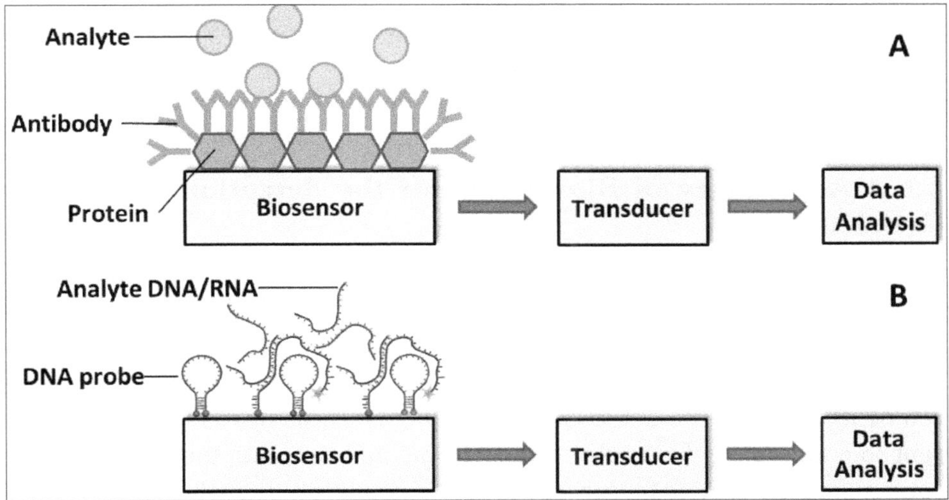

Figure 5.6 Schematic illustration of (A) antibody-based and (B) DNA/RNA-based biosensor for analyte detection. The specific combination of analyte and immobilized antibody (A) or DNA/RNA probe (B) produces a physicochemical change, such as mass, temperature, optical property or electrical potential. The change can be translated into a measurable signal for detection.8

5.5.2 Advantages of Early Detection of Plant Diseases

The use of biosensors for early detection of plant diseases has several advantages over traditional methods, such as visual inspection and laboratory analysis. Biosensors provide rapid and sensitive detection, allowing for early detection of plant diseases before visible symptoms appear. Additionally, biosensors can be used for on-site detection, which eliminates the need for samples to be sent to a laboratory for analysis, saving time and reducing the risk of sample contamination. Overall, biosensors have the potential to revolutionize the way plant diseases are detected and managed. By providing rapid, sensitive, and specific detection, biosensors can help farmers and plant pathologists to identify plant diseases early and take appropriate actions to prevent the spread of diseases and minimize crop losses.

5.6 Detection of Nutrient Deficiencies

Biosensors can be used for the detection of nutrient deficiencies in plants by measuring the concentration of specific nutrients in plant tissues. Nutrient deficiencies can have a significant impact on plant growth and crop yield, making early detection crucial for farmers and crop managers. Biosensors can be designed to detect specific nutrients, such as nitrogen, phosphorus, potassium, calcium, and magnesium, which are essential for plant growth. These biosensors can detect the concentration of nutrients in plant tissues, providing a rapid and sensitive method for the early detection of nutrient deficiencies. Some biosensors can detect multiple nutrients simultaneously, which can save time and reduce costs.

5.6.1 Advantages of Biosensors for the detection of Nutrient Deficiencies

The use of biosensors for the detection of nutrient deficiencies in plants has several advantages over traditional methods, such as soil analysis and visual inspection. Biosensors provide rapid and sensitive detection, allowing for early detection of nutrient deficiencies before visible symptoms appear. Additionally, biosensors can be used for on-site detection, which eliminates the need for samples to be sent to a laboratory for analysis, saving time and reducing the risk of sample contamination. Overall, biosensors have the potential to revolutionize the way nutrient deficiencies are detected and managed in plants. By providing rapid, sensitive, and specific detection, biosensors can help farmers and crop managers to identify nutrient deficiencies early and take appropriate actions to ensure optimal plant growth and crop yield.

5.7 Insect Pest Monitoring

Biosensors can be used for insect pest monitoring by detecting the specific biomolecules produced by insects, such as pheromones or enzymes. Biosensors can detect these biomolecules at very low concentrations, allowing for the early detection of insect pests before they cause significant damage to crops. Biosensors can be designed to detect specific insect pheromones, which are chemical compounds produced by insects for communication and mating purposes. By detecting these pheromones, biosensors can identify the presence of specific insect pests, allowing for targeted pest management strategies. Biosensors can also be designed to detect specific enzymes produced by insect pests, which are involved in the breakdown of plant tissues and the spread of diseases.

5.7.1 Advantages of Biosensors for Insect pest Monitoring

The use of biosensors for insect pest monitoring has several advantages over traditional methods, such as visual inspection and trapping. Biosensors provide rapid and sensitive detection, allowing for early detection of insect pests before significant damage occurs. Additionally, biosensors can be used for on-site detection, which eliminates the need for samples to be sent to a laboratory for analysis, saving time and reducing the risk of sample contamination. Overall, biosensors have the potential to revolutionize the way insect pests are monitored and managed in agriculture. By providing rapid, sensitive, and specific detection, biosensors can help farmers and crop managers to identify insect pests early and take appropriate actions to prevent crop damage and reduce the use of harmful pesticides.

5.8 Seed Quality Assessment

Biosensors can be used for seed quality assessment by detecting specific biomolecules or biochemical parameters that are associated with seed quality. Seed quality is critical for plant growth and crop yield, and the use of biosensors can provide a rapid and sensitive method for assessing seed quality.

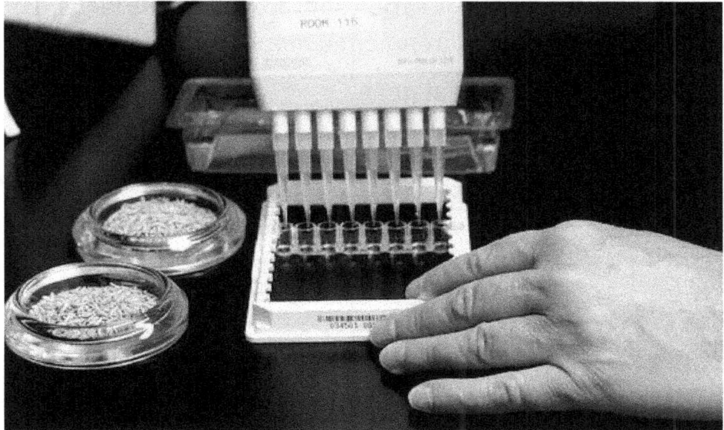

Figure 5.7 Schematic illustration of test to detect the presence of harmful pathogens in grass seed

Biosensors can be designed to detect various biomolecules or biochemical parameters associated with seed quality, such as germination rate, seed viability, seed vigor, and seed purity. For example, biosensors can detect enzymes or metabolites that are involved in seed germination, allowing for the rapid assessment of seed viability and vigor. Biosensors can also be used to detect contaminants or pathogens in seed samples, ensuring seed purity. The use of biosensors for seed quality assessment has several advantages over traditional methods, such as visual inspection and laboratory analysis. Biosensors provide rapid and sensitive detection, allowing for early detection of seed quality issues. Additionally, biosensors can be used for on-site detection, which eliminates the need for samples to be sent to a laboratory for analysis, saving time and reducing the risk of sample contamination.

Overall, biosensors have the potential to revolutionize the way seed quality is assessed in agriculture. By providing rapid, sensitive, and specific detection, biosensors can help farmers and seed producers to identify seed quality issues early and take appropriate actions to ensure optimal plant growth and crop yield.

5.9 E-nose (Electronic nose) Technology

E-nose (electronic nose) technology is a type of biosensor that mimics the olfactory system of animals, which is used to detect and identify odors and volatile compounds. E-nose technology uses an array of sensors that can detect and quantify volatile organic compounds (VOCs) in a sample, which can be used for a wide range of applications in agriculture asshown in Figure 5.8. In agriculture, e-nose technology can be used for various purposes, such as food quality control, plant disease diagnosis, and environmental monitoring.

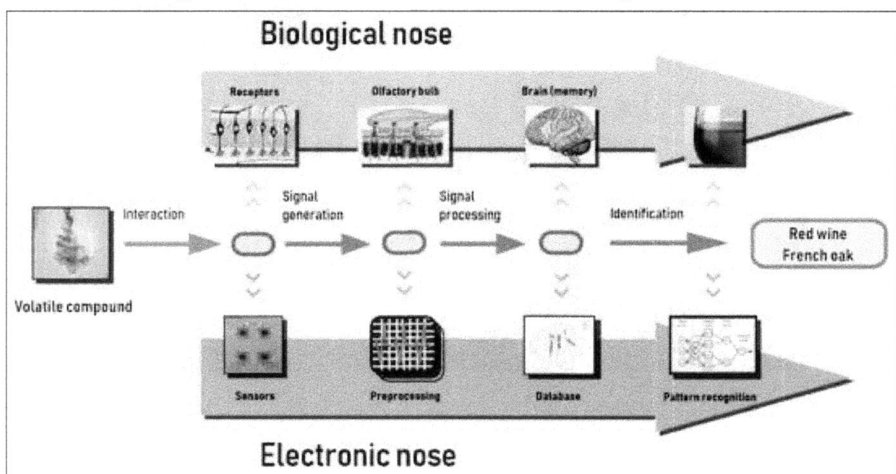

Figure 5.8 Schematic diagram representation that the test to detect the presence of harmful pathogens in grass seed.

For example, e-nose technology can be used to detect the presence of mycotoxins, which are toxic compounds produced by fungi that can contaminate crops and pose a risk to human and animal health. E-nose technology can also be used to detect and diagnose plant diseases by detecting specific VOCs produced by pathogenic microorganisms or infected plants. E-nose technology can also be used for environmental monitoring, such as detecting air pollutants, soil contaminants, and water quality. E-nose technology can detect the presence of specific VOCs in the environment, providing a rapid and sensitive method for monitoring pollution levels and assessing environmental health. The use of e-nose technology in agriculture has several advantages over traditional methods, such as laboratory analysis and visual inspection. E-nose technology provides rapid, sensitive, and specific detection, allowing for early detection and identification of various compounds and odors. Additionally, e-nose technology can be used for on-site detection, which eliminates the need for samples to be sent to a laboratory for analysis, saving time and reducing the risk of sample contamination.[13]

Overall, e-nose technology has the potential to revolutionize the way odors and volatile compounds are detected and identified in agriculture. By providing rapid, sensitive, and specific detection, e-nose technology can help farmers and agricultural professionals to identify various compounds and odors early and take appropriate actions to ensure optimal crop growth and environmental health.

5.10 Recent Advances in Micro-Nano Technologies

There have been several recent advances in micro-nano technologies that have the potential to revolutionize various fields, including agriculture. Here are some examples:

- **Nanosensor**: Nanosensor are biosensors that use nanomaterials to detect and quantify specific biomolecules or biochemical parameters. Recent advances in nanotechnology have led to the development of highly sensitive and selective nanosensor that can detect various targets, including pathogens, pesticides, and nutrients, at very low concentrations. nanosensor can be used for on-site detection, providing rapid and reliable results for various applications in agriculture, such as food safety, plant disease diagnosis, and nutrient monitoring.
- **Microfluidics:** Microfluidics is a technology that involves the manipulation of small volumes of fluids using microfabricated channels and chambers. Recent advances in microfluidics have led to the development of portable and low-cost devices that can perform various tasks, such as sample preparation, DNA amplification, and cell sorting, for various applications in agriculture. Microfluidic devices can be used for on-site detection, providing rapid and reliable results for various

applications in agriculture, such as plant disease diagnosis, food safety, and environmental monitoring.

- **Nanoparticles:** Nanoparticles are small particles with a size range of 1-100 nanometers, which have unique physicochemical properties that can be exploited for various applications. Recent advances in nanotechnology have led to the development of various nanoparticles, such as metal nanoparticles, magnetic nanoparticles, and quantum dots, that can be used for various applications in agriculture, such as pesticide delivery, plant growth promotion, and soil remediation.
- **3D printing:** 3D printing is a technology that involves the creation of three-dimensional objects by layer-by-layer deposition of materials. Recent advances in 3D printing have led to the development of various materials and devices that can be used for various applications in agriculture, such as soil sensors, irrigation systems, and plant growth substrates. 3D printing can be used to create customized and low-cost devices for various applications in agriculture, providing new opportunities for precision agriculture.

Overall, these recent advances in micro-nano technologies have the potential to transform various aspects of agriculture, from food safety and plant disease diagnosis to environmental monitoring and precision agriculture. These technologies provide new opportunities for on-site detection, precision farming, and sustainable agriculture, which can lead to improved crop yields, reduced environmental impact, and better food security.

5.11 Microfluidics and Lab-on-chip Technologies

Microfluidics and lab-on-chip technologies are two closely related fields that involve the manipulation and analysis of small volumes of fluids within microscale devices. These technologies have the potential to revolutionize various fields, including agriculture, by providing rapid and sensitive methods for sample preparation, analysis, and detection.

5.11.1 Microfluidics

Microfluidics involves the design and fabrication of microscale channels and chambers that can be used to manipulate and analyze small volumes of fluids, typically in the range of microliters to nanoliters. These devices can be used for various applications in agriculture, such as plant disease diagnosis, food safety, and environmental monitoring. Microfluidic devices are often portable and low-cost, making them ideal for on-site detection and analysis [15].

5.11.2 Lab-on-chip (LOC) Technologies

Lab-on-chip (LOC) technologies are a type of microfluidic device that integrates multiple functions, such as sample preparation, amplification, and detection, into a single device. LOC technologies can be used for various applications in agriculture, such as DNA analysis, pathogen detection, and nutrient monitoring as shown in Figure 5.9. LOC technologies are often used in combination with other technologies, such as PCR and biosensors, to provide rapid and sensitive detection of various targets.

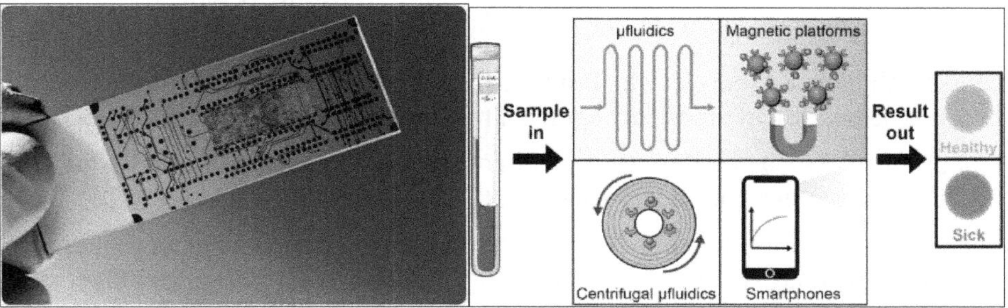

Figure 5.9 Photograph image representation that lab on chip and point care applications

5.11.3 Advantages of Microfluidics and Lab-on-chip Technologies

Both microfluidics and lab-on-chip technologies offer several advantages over traditional methods, such as reduced sample volume, faster analysis time, and improved sensitivity and specificity. Additionally, these technologies can be used for on-site detection, eliminating the need for samples to be sent to a laboratory for analysis, saving time and reducing the risk of sample contamination. Overall, microfluidics and lab-on-chip technologies have the potential to transform various aspects of agriculture, from food safety and plant disease diagnosis to environmental monitoring and precision agriculture. These technologies provide new opportunities for on-site detection, precision farming, and sustainable agriculture, which can lead to improved crop yields, reduced environmental impact, and better food security.

References

Kundu, Monika, et al. "Recent developments in biosensors to combat agricultural challenges and their future prospects." Trends in food sci. & tech. 88 (2019): 157-178.

McLamore, Eric S., et al. "FEAST of biosensors: Food, environmental and agricultural sensing technologies (FEAST) in North America." Biosensors and bioelectronics 178 (2021): 113011.

Sparks, Donald L., et al., eds. Methods of soil analysis, part 3: Chemical methods. Vol. 14. John Wiley & Sons, 2020.

Richardson, Susan D., and Susana Y. Kimura. "Water analysis: emerging contaminants and current issues." Analytical Chemistry 92.1 (2019): 473-505.

Umapathi, Reddicherla, et al. "Portable electrochemical sensing methodologies for on-site detection of pesticide residues in fruits and vegetables." Coordination Chemistry Reviews 453 (2022): 214305.

Pérez-Fernández, Beatriz, et al. "Electrochemical (Bio)Sensors for Pesticides Detection Using Screen-Printed Electrodes." Biosensors, vol. 10, no. 4, Apr. 2020, p. 32. Crossref,

Dyussembayev, Kazbek, et al. "Biosensor technologies for early detection and quantification of plant pathogens." Frontiers in Chemistry 9 (2021): 636245.

Fang, Yi, and Ramaraja Ramasamy. "Current and Prospective Methods for Plant Disease Detection." Biosensors, vol. 5, Aug. 2015, 537–61.

Barbedo, Jayme Garcia Arnal. "Detection of nutrition deficiencies in plants using proximal images and machine learning: A review." Computers and Electronics in Agriculture 162 (2019): 482-492.

Abd El-Ghany, Nesreen M., Shadia E. Abd El-Aziz, S. Marei. "A review: application of remote sensing as a promising strategy for insect pests and diseases management." Environmental Science and Pollution Research 27 (2020): 33503-33515.

ElMasry, Gamal, et al. "Recent applications of multispectral imaging in seed phenotyping and quality monitoring—An overview." Sensors 19.5 (2019): 1090.

Sánchez, Carlos, et al. "Use of Electronic Noses for Diagnosis of Digestive and Respiratory Diseases through the Breath." Biosensors, vol. 9, no. 1, Feb. 2019, p. 35.

Rasekh, Mansour, et al. "Classification and identification of essential oils from herbs and fruits based on a MOS electronic-nose technology." Chemosensors 9.6 (2021): 142.

Cheng, G., et al. "Improving micro-fine mineral flotation via micro/nano technologies." Separation Science and Technology 58.3 (2023): 520-537.

Battat, Sarah, David A. Weitz, and George M. Whitesides. "An outlook on microfluidics: the promise and the challenge." Lab on a Chip 22.3 (2022): 530-536.

Zhu, Hanliang, et al. "Recent advances in lab-on-a-chip technologies for viral diagnosis." Biosensors and Bioelectronics 153 (2020): 112041.

INDEX

www.ingramcontent.com/pod-product-compliance
Ingram Content Group UK Ltd.
Pitfield, Milton Keynes, MK11 3LW, UK
UKHW021956270726
14060UKWH00002B/546